GEORGIAN FURNITURE & LOOKING GLASSES

Frontispiece
DINING ROOM AT CARTON, COUNTY KILDARE, RESIDENCE OF LORD AND LADY BROCKET.
The multipedestal table recorded as "Ordered from Dublin."
Photograph courtesy of the late Noel C. Hartnell, Dublin.

GEORGIAN FURNITURE & LOOKING GLASSES

BY
F. LEWIS HINCKLEY

WITH 269 ILLUSTRATIONS

A WASHINGTON MEWS BOOK
Washington Square, New York

Other books by F. LEWIS HINCKLEY

Directory of the Historic Cabinet Woods

Directory of Queen Anne, Early Georgian & Chippendale Furniture: Establishing the Preeminence of the Dublin Craftsmen

Hepplewhite, Sheraton & Regency Furniture

Queen Anne & Georgian Looking Glasses

Metropolitan Furniture of the Georgian Years

The More Significant Georgian Furniture

Masterpieces of Queen Anne & Georgian Furniture

The More Significant Regency Furniture: 1800-1830+

Library of Congress Cataloging-in-Publication Data

Hinckley, F. Lewis.
Georgian furniture & looking glasses / by F. Lewis Hinckley.
p. cm.
Includes bibliographical references and index.
ISBN 0-8147-3486-3
1. Furniture, Georgian—Ireland—Dublin—Catalogs. 2. Furniture—Ireland—Dublin—History—18th century—Catalogs. 3. Mirrors—Ireland—Dublin—History—18th century—Catalogs. 4. Decoration and ornament—Ireland—Dublin—Georgian style—Catalogs. I. Title. II. Title: Georgian furniture and looking glasses.
NK2538.D8H552 1992
749.22'91835'09033—dc20 91-38288
CIP

Contents

Preface

WITH THE EXCEPTION of furniture shown in the frontispiece and last plate, all of the examples illustrated here have been hitherto described either as *English* or *American.* Just as they were actually produced, instead, in the long-ignored capital city of Dublin, the designs of American furniture will also be found to have been influenced not so much by London-trained craftsmen or London designs as they were through training acquired in Dublin, and designs originated there, as well as by the sister capital's adaptations of a very few patterns that were published in the London design books.

The existing confusion in museum labelings of such examples as *English* and even as *American,* and in the literature of the past seventy-five years, has resulted from a lack of scientific attention on the parts of those officials and of lay writers entrusted with informing collectors and the public in general. All have failed to realize that Georgian furniture and looking glasses of *metropolitan* quality in design, materiel, and tectonic skills, were produced solely in one or the other of the two great British capital cities: in London as the second largest city of the Western world during those years; or in Dublin, then third in size as well as in commercial and social importance. There were no other metropolitan furniture centers in England or in Ireland. Nevertheless, except for a relatively small number of easily identified London productions, all other fine and equally distinctive Georgian masterpieces have been misrepresented as *English* or as *American* ever since the collecting of such antiques came into vogue during the latter half of the nineteenth century.

Georgian Furniture and Looking Glasses

LITTLE IS KNOWN in regard to authentic London furniture and looking glasses because of scarcities that are indeed striking in comparison with the productions of other large cities of Europe and the British Isles. Of Continental cities only Paris was larger, while Dublin ranked next in size immediately after London, a relationship still to be discovered by British museum officials and equally uninformed literary authorities on the subject of so-called Old English Furniture. As a result of that scarcity, the subject of London furniture per se has been pursued only in occasional magazine articles rather than in books.

Obviously those articles have yet to be collated in museum files so that their safely documented illustrations may better serve toward a more comprehensive understanding of the designs that were actually produced in London itself. Such a systematic arrangement of consecutive developments in the English capital city might then have led to the first scientific investigation of high-quality British furniture as a whole, and thus to a realization of Dublin's major importance in its supply.

Equally shallow approaches to the entire subject are obvious in the first attempt to produce a modestly illustrated book supposedly concentrated on work actually carried out by the very few known London craftsmen. *Georgian Cabinet-Makers* was compiled by Ralph Edwards, former head keeper of the Victoria and Albert Museum, in collaboration with Margaret Jourdain, who had entrée to old Irish residences such as those pictured in *Country Life,* along with their Dublin, Parisian, and other Continental furnishings but, as first explained by R. W. Symonds, into which London furniture was never received. Despite the combined expertise of these two leading English authorities, a large proportion of the pieces they selected to illustrate that undertaking is in no way representative of London designs, but instead is entirely characteristic of those so abundantly produced in the sister capital of Dublin.

London furniture reliably identified by makers' labels is of greater rarity than that confirmed by bills, inventories, or other dependable documentation. This should have been evident from the inclusion, in a later publication, of only some half-dozen labeled pieces of which Symonds was able to provide photographs for illustrating Sir Ambrose Heale's book on his collection of trade cards, and his equally inconsequential listing of unknown London carpenters and other woodworkers taken from old street *Directories.*

Also, owing to the remarkable scarcity of genuine London furniture, the majority of Queen Anne and Georgian exhibits in British and American museums are labeled as

English, although any correctly determinable London example will, without exception, have been labeled as such. That great, but still unremarked overabundance of unexplainable *English* labels must eventually be questioned by an increasingly knowledgeable public, as to just where in the English hinterland such highly sophisticated metropolitan masterpieces might possibly have been designed and executed with such great creative ability and verve.

With the number of books and magazines that have been illustrated over the past seventy-five years by intermingling both Dublin and London masterpieces under the single erroneous classification of *Old English Furniture,* it is indeed high time for the two entirely different schools of design to be separately and scientifically studied and compared, so that the work of each widely separated metropolitan center of production can at long last be readily distinguished and appreciated for its own individual merits.

That aim will not be achieved as long as the assumed expertise of museum staffs does not permit consultations in depth with the more knowledgeable (and ethical) members of the British Antique Dealers' Association. These professionals cannot afford to oppose the defective system of classification to which they have been forced to adhere, after its undiscerning acceptance and persistent maintenance by all British and American museums ever since the earliest traders in *Old English Furniture* failed to disclose the actual and solitary source of their supplies. In the meantime these particular researches continue as the only ones thus far devoted to illustrating authentic specimens of Queen Anne and Georgian furniture and looking glasses as originated in the carelessly disregarded capital city of Dublin.

A projected multivolume presentation under the present title has been abandoned as cumbersome and unjustifiable, since American collectors of imported Queen Anne and Georgian masterpieces are not especially interested in which Dublin designs or patterns, or which types of looking glasses, were adopted or received in Philadelphia, New York, Boston, or other American towns; while those particular designs and patterns or types of looking glasses will be recognizable to all veteran collectors of Americana without having to be unduly stressed in the present text. Photographs set aside for that unwieldy project have been condensed so as to show only a more readily assimilated proportion of those examples having special interest to collectors in both fields.

Of greater value than all of the texts thus far expended on the subject of *Old English Furniture,* which have been so largely invalidated through their accompanyng illustrations of Dublin examples, are two simple phrases or rules of thumb that have proved serviceable in the determination of Dublin versus London designs. The more widely applicable of these came to mind in resolving the absurdly contrived Channon hoax,[1] when it became increasingly clear that no matter how much typography was being squandered in vain attempts to substantiate such a conspicuous act of chicanery, always and unfailingly it will be found that *Designs Speak For Themselves.*

An earlier instance of false evidence being employed to suggest a large and valuable London production, and also the aptness of this phrase, is that in which a newspaper or other printed advertisement was applied to an Early Georgian armchair, Fig. 104, in Percy Macquoid's *Age of Mahogany.* The advertisement reads: "GILES GRENDY, In St. John's Square, Clerkenwill, LONDON, Makes and Sells all Sorts of Cabinet Goods, Chairs, and Glasses." The chair selected for this particular deception is of a characteristic Dublin design that in no way resembles any of the work known to have been produced by Grendy.

Apparently because of that illustration, the lion-mask paw-foot side chair (Ill. 27) has been attributed to the same supposed London maker, even though the design is quite different and the carving much sharper and bolder than that in the Macquoid illustration. The suite represented by this single chair was owned by Lord DeLisle and Dudley, and apparently descended to him with other Dublin heirlooms that have been illustrated in the literature. All of the chairs, settees, and card tables of that suite retain their typical Irish needlepoint coverings and top panels. *Vide* color plates, *Age of Mahogany* and *Dictionary of English Furniture.*

Also on the basis of such a single false label the *bonheur du jour,* (Ill. 35) has been claimed as a London masterpiece, this one supposedly made by one Hugh Grainger, a seventeenth-century maker of "Fashionable Household Goods at Reasonable rates." The quality of the interior lacquer decoration, the skills spent in shaping and veneering the entire lower section, including the unusual and cross-grained supports, and the incorporation of a good-sized mirror panel with which the specially hinged door is faced, provide ample proof that this choice piece could not have been made and sold as a reasonably priced household item. Nor could such a highly sophisticated and cleverly executed tour de force have been produced during the turn of the seventeenth century, or prior to Grainger's supposed death in 1706. The design itself again bespeaks its origin and declares that such a piece could not have been made until at least a quarter of a century later, and then certainly not in London, but only in the persistently ignored capital of Dublin. (A table matching the undersection of this supposedly London piece was illustrated in the Sotheby Parke-Bernet catalogue of *The Benjamin Sonnenberg Collection,* Volume 2, lot 1674, as *Dutch circa 1720. Cf.* also my *Directory of the Historic Cabinet Woods,* 112.)

The second phrase that has proved helpful to these particular researches was voiced by Karl Freund, one of the earlier American traders to discover the wealth of Queen Anne and Georgian furniture to be found in Dublin and elsewhere throughout Ireland, when no supplies were available in London. I had received no hint of that ongoing trade during my own early years in the antique business, nor in later searches through Holland, North Germany, Denmark, and Sweden in hope of discovering a source for all those pieces of *Old English Furniture* that did not coincide with my own ideas of English traditions of design—regardless of their unquestioning acceptances in that light by all museum authorities and equally unprofessional lay writers.

Soon after those travels had yielded nothing of importance to my special interest I catalogued the contents of a Long Island home for sale at auction. There I found the furniture to be similar to that for which I had been attempting to establish a source. In its entirety it had been supplied by Karl Freund, fortunately along with an Early Georgian staircase removed from an old Irish mansion. Fortunately, also, I was able to reach Mr. Freund by phone to inquire if, while he was in Ireland, he had also bought furniture there. His answer was to exclaim: "What *IMAGINATION* those fellows had!!!"—a phrase that came regularly to mind from then on during the researches thus finally directed to the real source of such a large proportion of *Old English Furniture.* The *bonheur du jour,* (Ill. 35) so unsuitably associated with the name of a London maker is not only proof that *Designs Speak For Themselves,* it also exemplifies the originality, creative ability, and *joie de vivre* of Dublin craftsmen and so strikingly indicates the versatility and *Imagination* of their creations.

A knowledge of Dublin designs is desirable for collectors of Americana so that their influence on the development of American furniture can be recognized and appreciated;

and also because there is always the possibility of misunderstandings arising over pieces originating in a leading furniture center and those produced elsewhere that have been strongly influenced by its designs, patterns, or ornamental treatments. That knowledge can be of great value when Dublin pieces are offered for sale as genuine American productions. Prices under such circumstances have often reached figures ten times more, or even higher, than would have been obtained had the same pieces been routinely offered and accepted as *English.*

As an instance of such chicanery, the upholstered armchair (Ill. 78) is a typical Dublin example, as indicated by the carving of the arm terminals and supports, the seat edging, and especially the combination of blind-fret elements on the front legs. (These appear with greater clarity on the pilasters of a Dublin wardrobe illustrated, Fig. 144, in R. W. Symonds's *English Furniture From Charles II To George II.*) This chair was supposed to have been owned by George Washington, and also by William Penn (d. 1718!), and came from two successive generations of the Brooke family in Birdsboro, Pennsylvania. Some years ago it was sold as a "Rare Chippendale Mahogany Marlborough Armchair By Thomas Affleck of Philadelphia." It brought a record auction price at the time of $8,000, 1,000 per cent more than it would have brought as *English,* and equivalent to some $30,000 at least at present.

In some instances pieces of nondescript design also have been mistakenly acquired as Americana, such as a Queen Anne walnut wing chair with plain cabriole legs that appeared in a collection of Americana assembled by a famous lawyer and arbitrator who accompanied me during my appraisal of it. Pointing this one chair out as a foreign piece, he differed with my opinion, so I suggested that a thin cutting of the beech in its frame be submitted to the Forestry Products Laboratory in Madison, Wisconsin. In doing so he may have expressed his own opinion, recognized as "prompting" where second opinions are sought, for shortly thereafter he informed me that the cutting has been accepted as American beech. As the only instance of having to do so, I then asked that the cutting be microscopically reexamined, and an apology was sent the collector along with the information that through a recently discovered method of determination it was now possible to differentiate between American beech and the European variety represented by the cutting he had submitted.

Many handsome but nondescript examples of British and American seat furniture would be difficult or impossible to classify, even so broadly, were it not for their secondary woods. In particular this is true of wing chairs with Chippendale or later types of straight legs, and of upholstered open-arm easy or Marlborough chairs with similar underframes and plain incurvate arm supports. An important affirmation of Dublin's influence on American designs would be the discovery of a Marlborough armchair with the distinctively shaped and scroll-carved arm support introduced here through a Dublin pattern apparently used by Joseph Short of Newburyport, Massachusetts, in fashioning his lolling chairs of Martha Washington type, and smaller chairs with wood arms. In further confirmation of the designs, patterns, and ornamental treatments introduced in American cities and towns by Dublin-trained craftsmen, are ladder-back chairs with that same distinctive arm support and cross slats shaped and pierced as in Illustration 65. In fact, high-quality ladder-back chairs were not shown in any of the London design books and apparently were not produced there, for all those of fine metropolitan designs appearing in the London market over the past seventy-five years have been scientifically determinable as originating in Dublin, where a multiplicity of such models were produced during the middle and later Georgian years.

Contrasting with those plain and often undistinguishable armchairs are others of plain, simply molded frames, but which are fashioned with graceful oval or shield-shaped backs and serpentine cabriole legs through which they may be recognized as having originated in Dublin. The lightly carved example (Ill. 111), as well as those in Illustrations 108 to 110, were formerly accepted as *English* productions in the French-Hepplewhite style. In comparison with Parisian and other French or Continental *fauteuils*, however, it will be seen that they indicate a general transitional movement from rococo to classic designs, a movement more actively followed in Dublin than in London. In fact, one of the more angular examples, first considered as possibly a Potsdam production, was identified only after a console table *en suite* appeared next to it in these comparative proving files with the top surface featuring a typical Dublin painted decoration. London interpretations of this transitional trend differ conspicuously from those of Dublin; with a vague rendition of a simple French foot, as properly executed in the bergere (Ill. 110), shown in Hepplewhite's *Guide.*

An unusual reaction to this particular class of Dublin designs is that there is less hesitancy in describing the most outstandingly elegant examples as *beautiful*, rather than repeating the more customary superlative of *handsome*, as considered more appropriate to varying London masterpieces of exceptional excellence. This may be more acceptable from the fact that "the serpentine line of beauty" is a principal feature not only of these particular chairs, but also of those corresponding tables, commodes, and cabinets so handsomely treated in their differing types of ornamentation; *vide* Ills. 112, 123-125, 127, 129.

There has never been an illustrated survey of London seat furniture to give even so concise a representation of the designs actually carried out in that metropolis, as those covered even in this cursory presentation of Dublin seat furniture and cabinetwork. Indeed, while there is no marked scarcity of Dublin armchairs here or in any former publication of these researches, a French visitor to the English capital during the later years of the eighteenth century was astonished to find that "Londoners have scarcely any arm-chairs . . . in their apartments . . . they are satisfied with the common ones." As if to alleviate that scarcity, Dublin armchairs have been continuously illustrated as *English*, not only by the usual literary authorities, but also in many museum publications. Carrying the latter misrepresentations even further, attention has been drawn to a pair of heart-and-shield-back armchairs carved with *shamrocks* (one of which is shown here, Ill. 121) in the *Bulletin of the Metropolitan Museum of Art*, Vol. 24, 1929:

> The shield-back which Hepplewhite did so much to popularize, is further illustrated in a pair of armchairs . . . said to have come out of Ireland, but this should not be taken as conclusive proof of their Irish origin, it being an established fact that much of the finer furniture used in Ireland was made in England.

Thus additionally with museum endorsement a false and entirely imaginary assertion is put forth as an *established* fact.

An instance of other liberties taken and suggested by museum officials was that in which a correspondence with a photo section was interrupted by requests from the museum's inexperienced assistant keeper, and "expert on Irish furniture," begging me not to attribute some mysterious table to Robert Adam. Such requests were certainly presumptuous, as well as being so completely incomprehensible that I was moved to inquire just what had prompted them. My reply, from a full-fledged keeper, informed me that "there is absolutely no evidence that the table you wish to illustrate has anything to do with Robert Adam." With equal presumptuousness he suggested that "there is a table at Holkham of this type that can be attributed[2] to Benjamin Goodison with good reason."

After my business with the photo section had been completed, and a weak apology had finally been offered for an improbable mistake made by the junior official, I asked for a photograph of the mysterious table so that at least I might be able to recognize the piece over which so much fuss had been made. Instead, a print giving a wide view of a gallery displaying various types of furniture was sent. In this I could make out a table, but one that surely could not have come within the ken of the Scottish designer Robert Adam, nor that of the great London cabinetmaker Benjamin Goodison. Therefore a further inquiry was required to determine whether or not this unpretentious piece could possibly be the table of so much ado. That inquiry was replied to by still another individual, who assured me that such was the case. As an additional enigma, a separate print of that particular table had been easily available to fulfill my request, for such a print had been contained in my proving files for some years. It is reproduced here (Ill. 32), not only to show a typical Dublin example in every respect, but also to indicate the influence of such subsidiary designs on those of American furniture, in particular on Philadelphia card tables of *circa* 1750-1775, with similar legs, claw-and-ball feet and even balanced corneous or hornlike apron pendants.

Following the first acceptances, in museum and literary circles, of all high-quality Queen Anne and Georgian furniture as *English,* in accordance with that tongue-in-cheek designation by the American and English importers of such heirlooms obtained from ancestral homes in Ireland[3], the continuing confusion was augmented through the efforts of "experts" on Irish furniture. Magazine articles by such English authorities as R. W. Symonds and Margaret Jourdain portrayed Irish furniture in general as provincial in character or grotesque in design. *Vide Country Life,* March 14, 1952, "Furniture-Making in 18th-Century Ireland," by R. W. Symonds.

It was not until the closing years of his long career as the leading expert on *English* furniture that Symonds discovered, through searches made of Irish shipping records, that no old English furniture had ever been received in that country. Subsequently, in another magazine article, he admitted that "if a piece is known to have come out of Ireland it is accepted as having been made there." However, he was never able to identify any of the pieces not made known to him as coming out of Ireland by the importing dealers who provided them for so many of his illustrations. For as long as he continued in his authoritative capacity he continued to show his partiality for such Dublin masterpieces without ever becoming aware of that capital's existence as the long-overlooked source of their supply.

In the same *Country Life* article he mused:

> The question arises: what has happened to all the Irish oak, walnut, Japan and gilt furniture which was made, as the advertisements show, during the first 60 or so years of the 18th century?

The answer he never discovered is that along with an even greater number of examples in mahogany, yew, laburnum, padauk, sabicu, and other exotics, such *Irish* furniture, as produced in the capital city of Dublin, was already represented in many private collections such as that assembled by Percival Griffiths, on which Symonds drew so extensively for his illustrations; and in all leading British and American museums where they were, and still are, labeled as *English*.

The title of *expert on Irish furniture* is in itself an acknowledgment of a very limited acquaintance in that broad field, since it customarily excludes all knowledge of that country's capital-city masterpieces. Thus remaining completely ignorant of the characteristic designs produced there, such experts have been unable to distinguish between Dublin masterpieces and those produced in London. It follows that throughout the past hundred years the only authentic experts on Irish furniture have been those professional antique dealers who, through private and business connections in that country, have constantly supplied the British and American markets with the very finest of Queen Anne and Georgian furniture and looking glasses. By the same token these professionals are also the only bona fide experts on English furniture, rather than the unprofessional museum and literary experts who have so continuously proved themselves incapable of distinguishing between Dublin and London designs, of late even accepting as London productions the so-called Channon pieces (on the basis of false metal labelings) and the so-called Weekes's Museum pieces (on the basis of a misleading clock-dial inscription or inscriptions).

The Victoria and Albert Museum's picture book of *Georgian Furniture*, with Foreword by John Pope-Hennessy, Director (London, 1969), of which the first edition was the work of Ralph Edwards, was prepared by their "expert on Irish furniture." In addition to illustrating four of the recently discovered Dublin masterpieces as "attributed to the workshop of John Channon," this includes many others of equally characteristic Early Georgian, Chippendale, Hepplewhite, and Regency designs. Of those also illustrated here, the Dublin-Hepplewhite card table (Ill. 112) should have been obvious to any *expert on Irish furniture* from the inverse gadrooning that edges the serpentine apron and legs, an especially noticeable and identifying departure from orthodox procedure in London.

Had there been any consultation with the more knowledgeable and practical members of the antique business there would have been no hesitancies in disclosing the origin of an unattractive Irish commode (*op. cit.* Fig. 31), with especially distinctive Dublin openwork handles. In lacking such practical judgment and business experience, that commode was included with a confession that "This piece presents furniture historians with something of a problem. It may be an early English attempt to imitate a French bombé chest of drawers. Alternatively, it may be of Dutch origin"—the usual mistake of an amateur in misjudging Dublin furniture.

In my talks with old-time professionals about their buying trips to Ireland, no mention was ever made of the capital city. It would appear to have been taken for granted that there would be various sources for the supplies they had found so widely distributed throughout that country. It was not until much later that I thought it more likely that the capital was the one and only source. Then through the study of population statistics relative to the English and Irish capitals, small towns, and villages, I was finally able to realize that no other metropolis existed in either country; the only subsidiary furniture center in England was Lancaster, home of the much touted Gillow brothers, where a population of 9,050 was not reached until about 1815.

It was then clear that all Queen Anne and Georgian furniture and looking glasses of metropolitan designs, materiel, and working techniques could have been produced only in one or the other of the two capital cities, either in Dublin or in London. More attention was also paid to the geographic and sociological relationships of the sister capitals. With London situated in the far south-eastern corner of Great Britain, this was a definite handicap in any seaborne or overland deliveries to western areas of England, Wales, and Scotland, often entailing ten days to two weeks or more of travel time, while such areas were more promptly and conveniently reached from Dublin's location in the very heart of the British Isles. Also, from the amount of Dublin furniture, tax-free looking glasses, and Waterford glass lighting fixtures originally delivered to old English residences lying within a hundred miles of the capital, it became obvious that such Irish productions were even more widely competitive with those of London.

Little is known of Dublin's former sociological and socioeconomic relations with England, Scotland, and Wales; and it is long forgotten that once she was celebrated as "the gayest capital in all Europe." Wealthy Britishers formerly traveled to that capital for their annual Dublin Seasons, following similar rounds of social activities in London. During those visits there would be opportunities for acquiring seat furniture and cabinetwork of the very latest designs, and correspondent looking glasses available in that market without the exorbitant excise taxes that would have been added to their costs in London.

The finest and most sophisticated of Chippendale designs were produced in Dublin even prior to the publication of the first *Director.* This is obvious in the furnishings supplied, *circa* 1750, to the Duke of Beaufort for his famous Chinese-Chippendale bedroom at Badminton House, in Gloucestershire. The bed itself (*op. cit. Georgian Furniture* Frontispiece) was accompanied by an overmantel mirror (*vide Age of Mahogany,* Fig. 8). However its dragon finials were repeated on another looking glass (*vide* my *Queen Anne & Georgian Looking Glasses,* 220) also mislabeled in the same museum, but in this particular instance misinterpreted as *German!*

That misinterpretation serves to stress the long-continued confusion in museum circles, as explained by a former Dublin curator: "There are no criteria to tell what is Irish and what is English." Still it is clear that the three *designs speak for themselves* if viewed professionally, or with elementary scientific observation and discrimination. Nevertheless, there are likely to be continuing opinions that the Badminton pieces were delivered the long way round from the workshops of Thomas Chippendale, or that they were supplied from London by the Linnells. When museum authorities and equally unprofessional literary experts finally agree on a specific metropolitan origin for these Badminton pieces, the mirror long exhibited in the Victoria and Albert Museum as *German,* and all other related capital-city productions, this will be a signal that members of the antique profession can at long last reclassify their own Dublin masterpieces accordingly.

Among the numerous examples of Georgian furniture that I have examined over the past fifty years I have never come across any with the label of a London firm priveleged to announce itself as a *Maker* or *Makers to His Majesty.* On the other hand I have found an originally applied label, on the backing of a typical Dublin architectural pier mirror, of Joshua Kearney, Carver, Gilder and Looking Glass Maker to His Majesty. According to John Teahan of the National Museum of Ireland, city records in Dublin show Kearney at 186 Great Britain Street from 1795 to around 1805, when he moved to 49 Henry Street and remained there until at least 1820. These dates are also of special interest, as such frames are routinely accepted as of English Adam designs and dated accordingly.

Another Dublin firm to be so honored by an English king was that of Mack, Williams, and Gibton. This was unknown to the editors of the *Dictionary of English Furniture*, in its revised edition. There a mahogany dumbwaiter is illustrated as labeled by that firm with its address given only as *39 Stafford Street*, and the piece is dated *circa* 1760. Again my sole inquiry was addressed to John Teahan, who found that John Mack had started the firm in 1785 at 188 Abbey Street. From 1794 to 1800 he also had an address at 39 Stafford Street. He worked from 1805 to 1810 at that address in partnership with Gibton. When Williams entered the firm it continued as Mack, Williams, and Gibton at 39 Stafford Street from 1815 to 1825—the period in which the dumbwaiter was made, over a half-century later than *circa* 1760!

This is typical of the mistakes that have been made when *circa* dates have been unwisely volunteered throughout the literature in respect to unrecognized Dublin masterpieces. The same error was repeated by Edwards and Jourdain in their *Georgian Cabinet-Makers* (1955). There the label of Mack, Williams, and Gibton is recorded as appearing on "Each of a pair of fine serpentine-fronted mahogany commodes, c. 1760, with elaborate brass rococo handles, formerly in the Samuel Courtauld Collection." Thus, these were *Late Chippendale* commodes made between 1815 and 1825, rather than *circa* 1760.

In both instances a mystery still remains as to how *His Majesty* (King George), had become acquainted with the accomplishments of these particular Dublin businesses. It would seem likely that both had carried out royal commissions, as the king could not have inspected their work *en situ* since he had never visited the Irish capital. (Falling just within the two sets of dates, George IV visited Dublin in 1821.) He must have known and valued their particular skills more highly than those of their London competitors, whom he had not so signally honored.

Smaller mysteries surround the names and addresses of Dublin firms whose productions appear in various royal households, some of which are even recorded as to the very dates of their acquisitions. Of these there is the set of three Regency brass-inlaid amboina sofa tables made in 1816 for Princess Charlotte (*vide* my *The More Significant Georgian Furniture*, 180); the inlaid satinwood bookcase made for Queen Charlotte (*vide Age of Satinwood*, Fig. 181); and the Dublin-Hepplewhite carved mahogany shield-back armchair acquired by Princess Elizabeth (*vide* Oliver Brackett, *English Furniture, Illustrated*, p. 236).

The working dates of Mack, Williams, and Gibton are especially important in confirming my theories regarding the long-extended popularity of Chippendale designs in the production of Dublin furniture throughout the Hepplewhite, Sheraton, and Regency years. This may be evident only in a relative lack of patination on exposed surfaces, or a similar lack of oxidation on the unfinished surfaces of secondary softwoods or hardwoods. Thus the Bramshill armchair (Ill. 71) is such a Late Chippendale example, determinable in part through its design, but also through a relative lack of aging, especially in the seat frame. At other times a Late Chippendale date is even more obvious in the use of tectonic methods that were not introduced until two or more decades after the principal Chippendale years. Thus the Late Chippendale commode with elaborate brass rococo handles (Ill. 97) cannot date from the primary or earlier Chippendale years, for the drawer sides are fitted with rabbeted wood strips for receiving the bottom boards, a method adopted at the turn of the eighteenth century and followed to some extent even throughout the Victorian years.

From certain examples of Dublin furniture and looking glasses that have been published as belonging to English establishments of various sorts, it would seem that they have served, wittingly or unwittingly, in suggesting original ownerships or gifts of national productions. Thus a typical Dublin bow-back Windsor armchair with scroll-bracketed underframe and pronouncedly stilted pad feet, matching these features of Illustration 73, has been published (Oliver Brackett, *English Furniture, Illustrated,* p. 214) as owned by the Worshipful Company of Curriers; and a Dublin overmantel mirror similar to Illustration 192, with landscape panel "painted in the Dutch [NB] manner" has been illustrated in the *Dictionary of English Furniture,* Fig. 43, as "From the Star and Garter Hotel, Kew Bridge."

Just as the well-known Windsor armchair "bequeathed in 1774 by Oliver Goldsmith to his friend, Dr. Hawes" has served in a documentary capacity toward proving additional London and other English Windsors; a Windsor armchair published in *Country Life* (May 4, 1912), as from Lambay Island, near the coast of County Dublin, has served as a basic example in these additional researches. The sequence of designs leading from that piece has included others of the type sometimes misleadingly referred to as *Somerset chairs* (*vide* Ill. 77), apparently from the convenience of their deliveries to that southwestern English county. Those designs have in turn led to a bow-back example with central Gothic splat, and then to the fully developed Gothic Windsor armchair with upper section enclosed by a pointed arch as in Illustration 76, which is representative of a set of four that was formerly supplied to Clyne Castle, in South Wales.

Later Windsor examples exemplify the outstanding ingenuity of even the lesser Irish craftsmen and a continuing unstinted use of yew, except where, as in general practice, ash or elm was used for their scooped seats. Examples such as Illustration 145, displaying the ever-popular Irish use of Prince-of-Wales plumes, have been imaginatively designated as *Mendlesham chairs,* giving the implication that this accomplished and fairly sizable production emanated from the tiny Suffolk parish and village which, even according to the latest census report, presently contains only 934 inhabitants.

One of the so-called Somerset chairs provided the link to a design shown in the *Dictionary of English Furniture,* Fig. 111, which in turn led to the Late Georgian and Early Victorian designs with more vigorously scroll-cut central splat, and increasingly heavy turned legs joined by a bowed stretcher. Such an extravagant use of yew in this great abundance of remaining high-back tavern chairs (*vide* Ill. 146) would not have been possible in England during the middle and later decades of the nineteenth century. According to the *Cabinet Maker's Assistant* (London, 1853):

> The importance attached to this wood, and the chief inducement for its cultivation in former times—the manufacture of bows—have now ceased for centuries. It has unfortunately fallen into undeserved neglect, and has not been extensively planted with a view to its use in the arts. At present it is not to be procured in sufficient quantity, to be made generally available for the larger articles of furniture.

Yewwood was always much more plentiful in Ireland than in England, with some trees reaching diameters of ten feet or more. In his *Country Life* article on "Furniture-Making in

18th-Century Ireland'' Symonds seemed unaware of the continuous esteem in which this beautiful native wood was held for use in producing the very finest masterpieces of furniture, and for architectural work:

> The Irish liked yew wood more than the English did. There must have been a considerable number of trees growing in Ireland. Mr. Colin Johnston Robb illustrated a letter about yew wood (*Country Life,* November 30, 1951) with a photograph of the yew staircase in Birr Castle, King's County.

That great appreciation is also evident through the choice examples of yew furniture illustrated in my *Directory of the Historic Cabinet Woods,* all of which could have been produced only in Dublin, especially the magnificent figured solid yewwood architectural bookcase, repeated here (Ill. 134), with its original brass wire grilles and distinctive bail handles, both retaining traces of water gilding.

In this presentation of the more collectable Georgian masterpieces of furniture and looking glasses, to avoid as many redundancies as possible all further observations are restricted to those pieces that cannot be adequately covered in the captions, such as those relative to the two parcel-gilded walnut secretaires (Ill. 40 and 41). The smaller of these was illustrated several times by Symonds, who also published a scale drawing to show its *English* [*sic*] constructional features. He described it as "a scrutoire to end all scrutoires,'' but this was before the larger example had a brief appearance in London. Symonds looked upon all such masterpieces as being delivered straight from old English collections into the London antique market, rather than arriving there in regular shipments from an entirely different country whose capital city he never recognized as having any bearing on the subject of his expertise.

The Earl of Charlemont (Charlemont House, Parnel Square, Dublin) owned an architectural cabinet almost matching the one illustrated here (Ill. 43). *Vide* M. Harris, *English Furniture,* p. 42, 64. Both have been described as "in the manner of William Kent,'' and also as "attributed to,'' or "in the style of William Vile.'' It is unlikely that either would ever have considered the use of inverse gadrooning beneath the narrow plinth strip of the glazed door. Nevertheless, and despite such idiosyncratic details, every important cabinet featuring such distinctive carved panel moldings has been associated with the name of William Vile during recent decades, regardless of other equally characteristic but unrecognized Dublin treatments that do not resemble his known work, or that of any other London cabinetmaker.

Any serious researches in Queen Anne and Georgian furniture should take into account the importance of handles and escutcheons in establishing or confirming the London or Dublin origins of pieces otherwise mainly or partially determinable only by their designs. The fact that no such attention has been paid to the patterns of these fittings as they appear on safely identified London examples, in order that comparisons might then have been made with others appearing on pieces merely guessed at as *English,* is largely responsible for the continuing oversights in respect to the numerous misinterpretations of Dublin cabinetwork.

Had such studies been carried out by English museum or other lay authorities, it would have been impossible for an unscrupulous dealer to have manipulated the entire staffs of two separate museums, and their directors, into accepting and even enlarging upon the absurd Channon hoax, in which typical Dublin library tables, commodes, and cabinets, with especially characteristic Dublin handles, escutcheons, and other ornamental ormolu mounts, have been claimed for this unknown provincial cabinetmaker fortuitously discovered to have later worked in London. *Vide* the Victoria and Albert picture book of *Georgian Furniture*, 1969, *passim.*

The subject has been treated with continuing importance in the publication of these researches. Here it is briefly raised in connection with some of the most easily recognizable of Dublin patterns, such as those appearing in both Illustration 87 and 88, which is the same as used for the Badminton book table (*Age of Mahogany*, Fig. 161). The patterns shown in Illustrations 94, 95 and 96 also appear in the wardrobe previously referred to (R. W. Symonds, *English Furniture From Charles II To George II*, Fig. 144). Other patterns that have served in tracing and confirming the designs of Dublin cabinetwork will be mentioned in the captions; and none of them will be found on pieces actually made in London.

If the same trickster who launched the Channon hoax had acquired the bombé commode (Ill. 89) on his rounds of private Irish collections, he might, with his utter contempt for the lack of professional experience in the museum officials with whom he dealt, have arranged for another metal plaque to be inscribed with a *Fecit* message in favor of some equivalent of John Channon. A somewhat similar example had been described at auction as *Flemish or North German.* The figure it brought had been so disappointing that I was asked to inspect this one, determine its provenance, and suggest a more favorable presentation.

The handles indicated its wild Irish origin even more clearly than its form, inlaid panels, carving, and marble top. I suggested that it be classified in the usual manner as *English*, and that a Note be added to its catalogue description drawing attention to the influence of Justin-Aurele Meissonier on the designs of Thomas Chippendale—who actually plagiarized Meissonier's designs and published them as his own. As a result of following that advice this particular Dublin eccentricity brought $6,500 in its sale, or the equivalent of some $25,000 to $30,000 in today's debased currency.

Several English writers joined in praising the Chippendale commode (Ill. 90) as a London masterpiece of the first rank. Soon after coming into the London antique market it went on view at the *Thomas Chippendale Exhibition* at Temple Newsam House. On that occasion Miss Margaret Jourdain was "convinced that this piece of furniture could be attributed quite reasonably to Chippendale himself." It was also described, as usual, as actually coming "from the workshop of Thomas Chippendale" and as exhibiting to perfection the quiet dignity of the finest English furniture of the eighteenth century . . . "an outstanding example of cabinet-making of the period." Knowing of my principal interest, the owning dealer had sent me a photograph, agreeing with my determination of its Dublin, rather than London origin. Another Dublin commode with inverted corbels is illustrated in the *Age of Mahogany*, Fig. 6.

A hundred years ago it was widely known to English society that the masterpieces now eulogized as coming from the workshop of this or that known or unknown London craftsman were to be obtained in the same manner as that followed by Lord St. Oswald in the refurbishing of Nostel Priory, by Lady Randolph Churchill in obtaining "choice pieces from ancestral homes," and by other members of the aristocracy seeking furnishings for their English homes—simply by searching for them in the Dublin antique shops. When American and English antique dealers discovered the same sources of supply, they also established their own private connections with owners of Irish castles and mansions. Thus collectors and museums have been able to form collections that would have been meager at best had they been entirely dependent on the seldom available London productions; for old English families have seldom parted with their heirlooms, while Irish families have been doing so ever since the collecting of *Old English Furniture* came into vogue.

Because of that great and continuing exodus from Ireland, designs of furniture and looking glasses formerly produced in the capital city go unrecognized in that country today, outside of trade circles. National museums admit only pieces of inferior quality as "probably Irish"; while all of their exhibits that have been published in the literature of the past seventy-five years are accepted as *English.* There has been no interest in keeping up with English investigations such as the findings of R. W. Symonds that English furniture was not received in old Irish residences, and that if a piece is known to have come out of Ireland it is accepted as having been produced there.

It came as no surprise, therefore, when the president of The Irish Georgian Society, Desmond Guinness, in collaboration with William Ryan, in 1971, published *Irish Houses and Castles,* with only three short paragraphs naïvely covering the entire subject of Irish furniture and its influence on American furniture designs:

> Today, works of art are more likely to leave Ireland than to be brought in, and there is unfortunately no law to prevent this. Antique dealers have for years now been scouting the country, buying Irish furniture to sell abroad. Some has been sold as "American"; both American and Irish furniture are provincial expressions of English, and Irish cabinet-makers are known to have set up shop in Philadelphia in the eighteenth century. Of late, however, research at Winterthur and elsewhere into the secondary woods has put identification on a more scientific basis.
>
> Irish furniture has a very definite character of its own. It is often black—even mahogany was sometimes blackened to resemble bog oak—and the carving on it is unlike anything made in England. The most interesting period, as with architecture, plaster-work, silver, and so on, is from 1725 to 1775. After that, the design was brought into line with English furniture of the Sheraton variety. The mask was widely used, both as a centerpiece and on the knees of side-tables. Whether lion, satyr, or human heads, the features are twisted into a spine-chilling grimace, very different from the expressionless heads on English furniture, apparently gazing out to sea and definitely tired of the view. The wide apron of the Irish side-table is a tumbling mass of exuberant carving; snakes eating their own tails, fruits, flowers, little distorted animal masks with haloes of shells. In its movement and madness this carving seems to conjure up the imaginary animal and human figures that decorate the pages of the Book of Kells.
>
> Oval hunting-tables with absolutely plain legs were used for dining;[4] the flaps could drop down when the table was taken outdoors for a lawn meet. They were also known as wake tables; at a wake, the coffin would be laid down the middle, surrounded by food and drink.

Even though the grand interior views of *Irish Houses and Castles* do not show their most significant furnishings with any great degree of clarity, they do in themselves sparingly cover the history of Dublin furniture, as well as Dublin or Waterford lighting fixtures. Had these illustrations been available *en masse* thirty years earlier they would have been a great help in facilitating the present researches. They are especially rich in the earlier styles of Dublin looking glasses, but include even those of Sheraton and Regency designs; and all are unmistakable documentary examples. The Chippendale mirrors are of the utmost importance to present-day scholars. One that was formerly at Tyrone House, Dublin, is shown, p. 331, as removed to Powerscourt.

Antique dealers mentioned as buying Irish furniture to sell abroad, even as *American*, are responsible to only a negligible degree for the long-continued lack of knowledge as to just what is Irish and what is *English*, or which are Dublin and which are London examples. In calling everything English, as they openly acknowledged when finally questioned about that trade, no real harm has resulted from purchases made under that designation, for unrecognized Dublin examples have steadily increased in aesthetic appreciation, and therefore in pecuniary values throughout the past hundred years. Adverse effects have been those resulting from the lack of scientific attention to the designs of these Dublin masterpieces vis-à-vis those of London, on the parts of museum authorities and equally dilettantish lay writers. The more knowledgeable and ethical antique dealers of today can eventually clear up the far greater deceptions culminating in museum acceptances of the so-called Weekes's Museum, Langlois, and Channon pieces, but only when they, too, are finally questioned about that trade.

The Chippendale overmantel mirror (Ill. 237) is reminiscent of a shrewd but strictly ethical early trader with whom I had friendly relations in winding up his antique business. At the time I was still unaware of the source of such items, and I queried him as to whether he considered the mirror likely to be North German or English. In suiting his reply to my wording he answered *English.* Had I been knowledgeable enough at the time to have said *Irish* or English, in his complete lack of duplicity he would have answered *Irish.* However, my question had made it clear that I was not cognizant of that covert trade, and it would have been both uncalled for and unbusinesslike for him to have disclosed his source of supply. Since I could not accept the mirror, with its summital shell motif, as English, I catalogued it as *American* with a note explaining that reaction; and its auction price suggested that it had been bought in that light.

The same dealer was perfectly amenable to have a table suggested as possibly of a Massachusetts origin, but he would not stoop to having the brass ball feet removed in order to lend credence to that suggestion. Thus in permitting the integrity of the table to remain untouched they were left in typification of its Dublin origin. The rest of the table is practically indistinguishable in its design from those two-part dining tables generally assigned to Massachusetts, the most famous of which is in the president's dining room of the White House.

Among the numerous books on old Irish and English glass, such as those by M. S. Dudley Westropp, Mrs. Graydon Stanus, and N. Hudson Moore, there are only bits and pieces to be gleaned pertinent to the glasswares supplied by Dublin, rather than those delivered from Waterford to Great Britain and America. It may be learned, however, that in addition to the manufacture of "plate glass for looking glasses, windows, and coach glasses," Dublin, as well as Waterford, made "glass lustres, girandoles, chandelieres, candelabra, etc."

Collectors of Americana will have become acquainted with coach and chariot glasses from the advertisements of John Elliott and his son, who could not have carried on their retail and wholesale business in these, and more particularly in the looking glasses that bear their labels, except by obtaining them, tax-free, directly from Dublin. No matter how many deliveries they received "in ships just arrived from London," the costs in that capital, plus the exorbitant excise duties that would have had to be added to those costs, would have made such commerce too unprofitable to be endured, had the same designs of mirror frames been made there.

Again, *designs* of the parcel-gilded looking glasses distributed by the Elliotts in Philadelphia and by William Wilmerding in New York prove not to have been either *American or English* but to have been produced in, and exported from Dublin, just as those of the more sophisticated Late Georgian styles indicate that they, too, were received from the same capital to be distributed by such importers as Stephen Whiting of Boston and James Stokes of Philadelphia. Additionally, as an exception to the usual trade with neighboring counties of Great Britain, eventually this was extended to the east coast of Scotland, where Dublin mirrors of Sheraton and Regency designs were stocked by a Looking Glass Store in Edinburgh, more convenient by far to the heavily taxed port of London.

Eventually it will be recognized by all collectors of American furniture that the different Queen Anne and Georgian designs of walnut or mahogany looking glasses with various decorative treatments, which to some extent were retailed in Philadelphia, Boston, and New York, are all unreservedly of Dublin, rather than of American or English origins.

Dublin furniture was influenced to varying extents by French, German, Danish, Dutch, and Italian designs and/or decorative treatments. Aside from the usual misrepresentations as *English*, and recently even as rare *London* extravaganzas, Dublin productions may also be mistakenly assigned to any of those countries, especially to Holland, or to the Continent as a whole. Thus when one of the earliest types of Dublin looking glasses[5] with arched frames exactly shaped and bordered as in Illustration 167, but in which the borders are treated with *verre églomizeé* decorations (*vide Dictionary of English Furniture*, Fig. 21) was believed by R. W. Symonds to be *Continental*, he failed as usual to recognize a favorite Dublin mirror form, with border panels obviously executed in the Irish capital by Venetian artists in that medium.

It is likely that Venetian craftsmen were more generally employed in Dublin and in partially completed Irish residences of outlying districts, than in London or elsewhere in Great Britain. The overdoor mosaic in the salon at Powerscourt, Enniskerry, County Wicklow (Plate 127) is recorded as having been executed by *Salviati of Venice*. While the name or names of his compatriots responsible for the sculpturesque pier mirrors and tables have not been recorded, they are representative of Venetian workmanship of the highest order.[6] Obviously they also were executed *in situ* rather than in Venice, and the exceptionally large mirror plates must, at great expense, have been specially made in and transported from the nearby capital city.

Another gilded looking glass in which the arched upper section is shaped exactly as in Illustration 167, and a red-and-gold églomisé border is also divided into five similarly spaced panels, features a very elaborate armorial cresting. This example from Penshurst was loaned by Lord DeLisle and Dudley to an exhibition of London masterpieces. Subsequently photographed and published in a magazine article, the halftone illustration included in these comparative proving files has become an especially important document, since it is now clear that unrecorded Dublin craftsmen supplied most if not all of that family's heirlooms as illustrated throughout the literature on *Old English Furniture* by Percy Macquoid, Macquoid and Edwards, et al.

It may be that Venetian or other Italian glassworkers have been responsible for cutting the conventional patterns that appear in many upper mirror plates of the most important Dublin gilded looking glasses, with or without églomisé panels, and also in those of parcel-gilded walnut or mahogany examples. Because of the simple star-and-sheaf cutting of Illustration 167 and its seemingly undeterminable form, additional proof had gone unnoticed until recent comparisons with similarly shaped and paneled borders of more importantly treated, and more imaginatively cut, églomisé examples have finally resulted in its acceptance as also originating in Dublin, and as possibly cut by an Italian glassworker.

While making those comparisons it was also realized that the photograph of this particular mirror had not been requested through seeing the piece itself, or through an advertisement of the dealer. Instead as a very thoughtful gesture it had been included in a selection made up by his widow, who had accompanied him on his buying trips abroad; obviously this included visits to the Dublin antique dealers, in regard to which I had never questioned either. The selection had also included a photograph of a rare tassel-back chair in which the arm supports flare just above the seat in bifurcations edged with inside scroll carvings. Outside of Dublin this feature was repeated only in Massachusetts seat furniture, possibly that made by Joseph Short of Newburyport.

Another example in which the arm terminals and supports are whorled together (*vide* Ill. 7) is one of a considerable number of similarly treated Dublin armchairs and settees already illustrated from these files (*vide* my *Directory of Queen Anne, Early Georgian and Chippendale Furniture*, 67-68, 72, 87). These whorled arm terminals served in identifying the solitary Dublin armchair that has been so naïvely accredited to the unknown work of John Channon. This is one imaginative feature of Dublin seat furniture that is not to be found in any other productions whatsoever, of Great Britain, America, or the European Continent.

One of the primary clues to Irish looking glasses of Chippendale designs is the use of stalagmites rising from various crest or lateral treatments, especially where the gilded frames are not of the first quality in their designs and carvings; *vide* Wallace Nutting, *Furniture Treasury*, 2997. Although these subsidiary examples do not seem to represent fine metropolitan work, one very similar to the example in Edgar G. Miller, *American Antique Furniture*, 1228, appears in Guinness and Ryan, *Irish Houses and Castles*, p. 297, as contained in Belevedere, Mullinger, County Westmeath, which lies within easy distance of the capital. Much finer Chippendale examples appear (*op. cit.* p. 79) in Kilshannig, County Cork, and (p. 317) in Mount Kennedy, County Wicklow. The authors have also failed to recognize a number of typical Dublin convex mirrors, including one (p. 90) in Glenveagh Castle, County Donegal, that is a mate to the Late Regency example illustrated here (Ill. 266) as well as to the one shown by them as appearing in Moore Abbey, County Kildare.

"A great Choice of guilt Leather and painted screens" was available to readers of the *Dublin Journal* during the 1740s. Contrasting with those displaying genre or pastoral subjects in soft rich colors against gold-scale grounds, supposedly copied from pictures by Lancret and other French artists, is the example here (Pl. 128). In this, balanced panels of floral urns enclose a terraced arrangement of Chinese buildings and figures; while in others similar or matching urns of flowers were used to surround exotic birds amid a profusion of blossoms and foliage. With these two particular types of Dublin screens the folding panels were edged and hinged with narrow strips of gilded leather, the inner hinge strips as well as the outside edging strips all studded with evenly spaced brass nail heads.

Another category of generally unrecognized Dublin artistry is represented here in Plate 126 by the supposedly *Louis XVI* chandeliere sold as *French XVIII Century* at $2,000—when that auction figure was equivalent to $8,000 in relation to today's debased currency. Similar designs, actually dating between *circa* 1810-1820, have also been published as *Dutch*, the usual surmise extending even to this particular type of Dublin production. A simpler version is illustrated in Margaret Jourdain, *Regency Furniture* (Fig. 200), as "From the Earl of Caledon, Caledon, Ireland"; and a more elaborate later example appears in Fig. 202.

1. *Vide* Victoria & Albert Museum *Bulletin*, Vol. 2, No. 2 (April 1966).

2. Such attributions by English lay authorities are in general optimistic attempts to associate London names with unrecognized or unacknowledged Dublin creations of fine or superfine capital-city quality. Hence they are not to be taken seriously.

3. "During her Ireland years (1877-1881) Lady Churchill was able to obtain choice pieces from ancestral homes in her rounds of the Dublin antique shops." Ralph G. Martin, *Jennie*, Vol. 1, 1959.

4. *Cf.* with the present frontispiece.

5. Other seemingly early, supposedly Charles II, looking glasses crudely carved with cherubs, foliage, etc. (*vide* R. W. Symonds, *Furniture-Making in 17th and 18th Century England*, Fig. 198) are actually Dublin productions dating from the later decades of the eighteenth century. These carvings are also approximated in the contemporary baroque stands of so-called Charles II lacquer cabinets decorated in the Chinese taste.

6. Venetian carving is also represented in the pier mirror formerly owned by the Duke of Kent, which has been acquired by Temple Newsam, Leeds.

Plates

1 EARLY GEORGIAN WALNUT ARMCHAIR.

2 EARLY GEORGIAN WALNUT ARMCHAIR.

3 EARLY GEORGIAN WALNUT ARMCHAIR. A design usually associated with the earlier, Queen Anne years.

4 EARLY GEORGIAN WALNUT ARMCHAIR.

5 EARLY GEORGIAN WALNUT SIDE CHAIR. With characteristic oak-leaf carving on the knees; *cf.* Ill. 21.

6 EARLY GEORGIAN WALNUT ARMCHAIR.

7 EARLY GEORGIAN WALNUT ARMCHAIR. The whorled-together arm terminals and supports, a most distinctive Dublin treatment, is not found in London, or any other, seat furniture.

8 EARLY GEORGIAN SHELL-CARVED WALNUT EAGLE-HEAD ARMCHAIR.

9 EARLY GEORGIAN WALNUT EAGLE-HEAD ARMCHAIR.

10 EARLY GEORGIAN WALNUT TASSEL-BACK SIDE CHAIR.

11 EARLY GEORGIAN-CHIPPENDALE WALNUT SIDE CHAIR. The *pretzel* splat, introduced in Philadelphia and New England, sometimes claimed as representing *Dutch* influence. An armchair of this general design, in the National Museum of Ireland, has been widely publicized as *English circa 1725-30.*

12 EARLY GEORGIAN MAHOGANY ARMCHAIR. The entire front leg reproduced in American seat furniture, lowboys, etc.

13 EARLY GEORGIAN MAHOGANY TASSEL-BACK SIDE CHAIR. A matching eagle-head armchair appeared in an exhibition and sale entitled *A Century of American Chairs. Vide* Macquoid, *Age of Mahogany,* Fig. 119.

14 EARLY GEORGIAN MAHOGANY ARMCHAIR. Courtesy of French & Company, Inc., New York City.

15 EARLY GEORGIAN MAHOGANY TASSEL-BACK SIDE CHAIR.

16 EARLY GEORGIAN MAHOGANY TASSEL-BACK CORNER CHAIR. First imported and sold as *English,* then misrepresented instead as a *Philadelphia* masterpiece.

17 EARLY GEORGIAN MAHOGANY *NEW YORK* SIDE CHAIR.
The tassel splat rising from a characteristic Dublin shaped shoe piece.

18 EARLY GEORGIAN MAHOGANY *NEW YORK* SIDE CHAIR.
Cf. knee carving of Ill. 25.

19 EARLY GEORGIAN MAHOGANY EAGLE-HEAD ARMCHAIR.

20 EARLY GEORGIAN MAHOGANY EAGLE-HEAD ARMCHAIR.

21 EARLY GEORGIAN MAHOGANY EAGLE-HEAD ARMCHAIR. With distinctive oak-leaf carving; *cf.* Ill. 5.

22 EARLY GEORGIAN MAHOGANY SIDE CHAIR. Described as *American or English.*

23 EARLY GEORGIAN MAHOGANY SIDE CHAIR. With frill-carved crest rail.

24 EARLY GEORGIAN MAHOGANY *NEW YORK* SIDE CHAIR. A Dublin design also found in Scandinavian collections.

25 EARLY GEORGIAN PAW-FOOT SIDE CHAIR. The knee carving similar to that appearing in New York card tables of Dublin designs.
Courtesy of Maple & Co. Ltd., London.

26 EARLY GEORGIAN WALNUT WING CHAIR. The trifid or drake foot introduced in American furniture by Dublin-trained craftsmen.

27 EARLY GEORGIAN MAHOGANY SIDE CHAIR WITH ORIGINAL IRISH NEEDLEPOINT SEAT AND BACK PANELS. Collection of Lord DeLisle and Dudley, Copped Hall, Essex, and Penshurst Place, Kent.
Courtesy of Leicester Museums and Art Gallery.

28 EARLY GEORGIAN WALNUT WING CHAIR IN ORIGINAL IRISH NEEDLEPOINT COVERING. The trifid or drake foot adopted in American furniture.

29 EARLY GEORGIAN WALNUT TABLE. Opening with a concertina action.
Courtesy of French & Company, Inc., New York City.

30 EARLY GEORGIAN WALNUT CARD TABLE.

31 EARLY GEORGIAN MAHOGANY CARD TABLE.

32 EARLY GEORGIAN MAHOGANY CARD TABLE. Reflecting Dublin's influence, especially in Philadelphia furniture. *Vide p. 17.*

33 EARLY GEORGIAN DECORATED LACQUER DRESSING MIRROR.

34 EARLY GEORGIAN WALNUT AND BURL WALNUT BUREAU-ON-STAND. A matching piece surmounted by a dressing mirror is illustrated in R. W. Symonds, *English Furniture From Charles II To George II*, Fig. 84.

35 GEORGE II WALNUT BONHEUR DU JOUR WITH DECORATED LACQUER INTERIOR. The spirited elegance, the *joie de vivre* of this imaginative creation has never been equaled in even the finest of London productions.
Courtesy of Christie, Manson, & Woods, London.

36 EARLY GEORGIAN WALNUT ARCHITECTURAL CHEST-ON-STAND. With ivory claws. The knee edgings similar to those appearing in subsidiary types of Dublin seat furniture and cabinetwork. *Cf.* Ill. 44.
Victoria & Albert Museum. Crown Copyright.

37 EARLY GEORGIAN WALNUT CHEST-ON-CHEST. Victoria & Albert Museum. Crown Copyright.

38 EARLY GEORGIAN WALNUT CHEST-ON-CHEST WITH WRITING DRAWER.

39 EARLY GEORGIAN WALNUT SECRETAIRE. Courtesy of the Nordiska Museet, Stockholm.

40 EARLY GEORGIAN PARCEL-GILDED WALNUT ARCHITECTURAL SECRETAIRE.

41 EARLY GEORGIAN PARCEL-GILDED WALNUT ARCHITECTURAL SECRETAIRE.

42 EARLY GEORGIAN MAHOGANY ARCHITECTURAL SECRETAIRE.

43 EARLY GEORGIAN MAHOGANY ARCHITECTURAL BREAKFRONT CABINET.
Described as the *work of a first-class London cabinet-maker*; as *by William Vile*, etc.
Victoria & Albert Museum. Crown Copyright.

44 EARLY GEORGIAN-CHIPPENDALE MAHOGANY ARMCHAIR. *Cf.* central crest motif with centers of New England block-front shell carvings.

45 EARLY GEORGIAN-CHIPPENDALE MAHOGANY PAW-FOOT SIDE CHAIR.

46 CHIPPENDALE MAHOGANY ARMCHAIR.

47 CHIPPENDALE MAHOGANY EAGLE-HEAD ARMCHAIR. The splat pattern favored by Benjamin Randolph and other Philadelphia chair makers.

48 CHIPPENDALE FRET-CARVED MAHOGANY ARMCHAIR.

49 CHIPPENDALE MAHOGANY SIDE CHAIR.

50 CHIPPENDALE MAHOGANY SIDE CHAIR. *Cf.* Ill. 25.

51 CHIPPENDALE MAHOGANY SIDE CHAIR.

52 CHIPPENDALE MAHOGANY SIDE CHAIR.

53 CHIPPENDALE MAHOGANY SIDE CHAIR. The splat pattern copied in Massachusetts.

54 CHIPPENDALE MAHOGANY ARMCHAIR. *Cf.* Ills. 47 and 55.

55 CHIPPENDALE MAHOGANY ARMCHAIR. The entire back duplicated in Philadelphia chairs attributed to Benjamin Randolph.

56 CHIPPENDALE MAHOGANY ARMCHAIR.

57 CHIPPENDALE MAHOGANY ARMCHAIR.

58 CHIPPENDALE MAHOGANY SIDE CHAIR.

59 CHIPPENDALE MAHOGANY SIDE CHAIR.
Victoria & Albert Museum. Crown Copyright.

60 CHIPPENDALE MAHOGANY SIDE CHAIR.
The Gothic pattern of the back reproduced in New England seat furniture.

61 CHIPPENDALE MAHOGANY SIDE CHAIR. *Cf.* Ill. 57.
Victoria & Albert Museum. Crown Copyright.

62 CHIPPENDALE MAHOGANY SIDE CHAIR. Sold as *American*. The Gothic splat pattern duplicated in typical Philadelphia armchairs.

63 CHIPPENDALE WALNUT SIDE CHAIR.
The splat pattern followed in Philadelphia, and in chairs attributed to James Gillingham.

64 CHIPPENDALE MAHOGANY LADDER-BACK ARMCHAIR.

65 CHIPPENDALE MAHOGANY LADDER-BACK SIDE CHAIR. The serpentine pierced cross slats copied in Massachusetts ladder backs, some attributed to Joseph Short of Newburyport.

66 CHIPPENDALE MAHOGANY LADDER-BACK SIDE CHAIR.

67 CHIPPENDALE MAHOGANY LADDER-BACK ARMCHAIR.

68 CHINESE-CHIPPENDALE MAHOGANY ARMCHAIR. Designs like these were included in English collections through importations received prior to the turn of the nineteenth century.

69 CHINESE-CHIPPENDALE MAHOGANY SIDE CHAIR. Similar designs were made with eagle-head arm terminals.

70 LATE CHINESE-CHIPPENDALE MAHOGANY ARMCHAIR. From Bramshill, Hampshire.

71 LATE CHIPPENDALE MAHOGANY ARMCHAIR. The blocked feet intended to receive castors, in this instance of early nineteenth-century manufacture. From Bramshill, Hampshire.

72 CHIPPENDALE MAHOGANY HIGH-BACK CORNER READING CHAIR.

73 CHIPPENDALE YEWWOOD WINDSOR ARMCHAIR. Victoria & Albert Museum. Crown Copyright.

74 CHIPPENDALE MAHOGANY HIGH-BACK CORNER READING CHAIR.

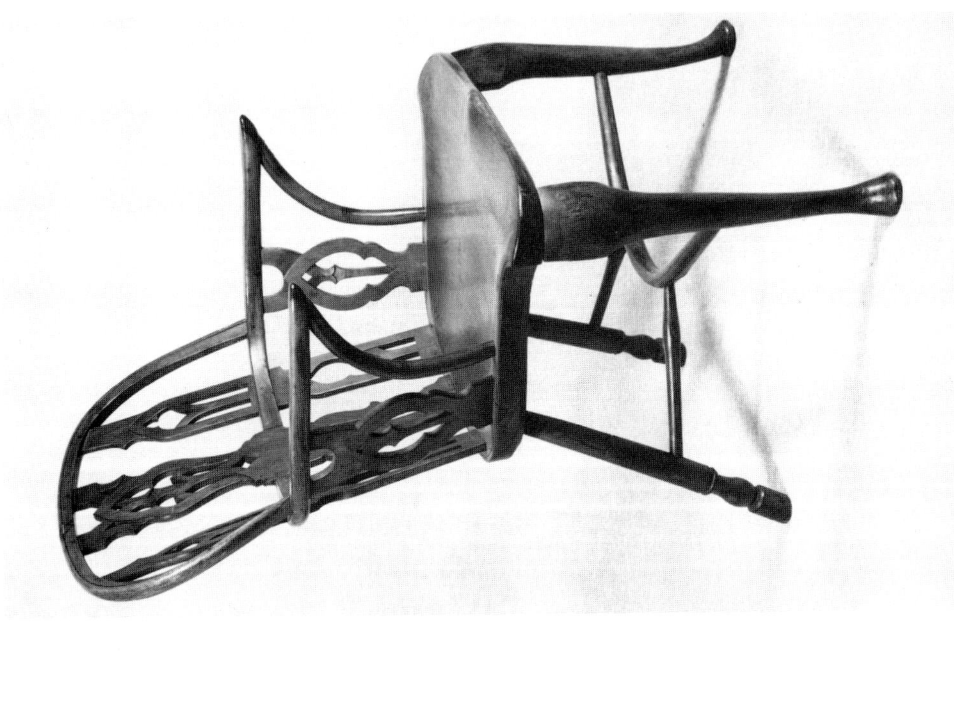

75 CHIPPENDALE YEWWOOD WINDSOR ARMCHAIR.
With turned rear legs matching those of Ill. 76.
Victoria & Albert Museum. Crown Copyright.

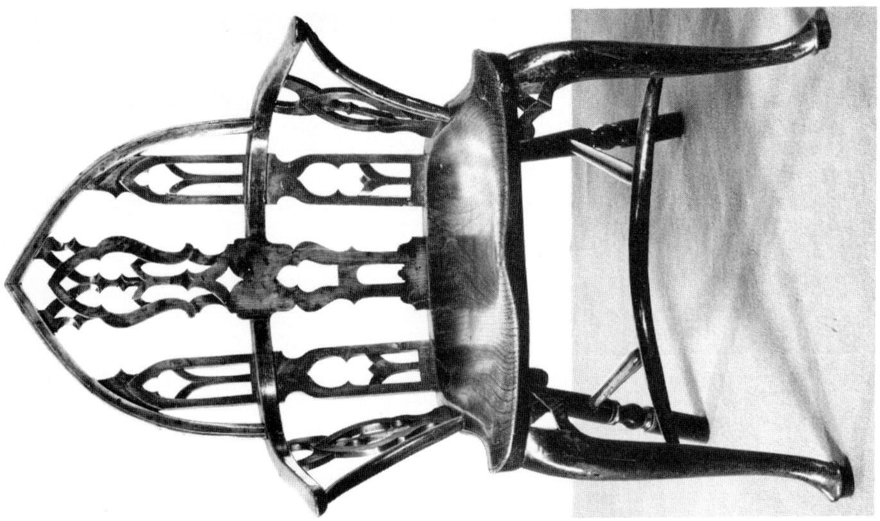

76 CHIPPENDALE YEWWOOD WINDSOR ARMCHAIR.
Courtesy of Mallett & Son (Antiques) Ltd.,
London and Geneva.

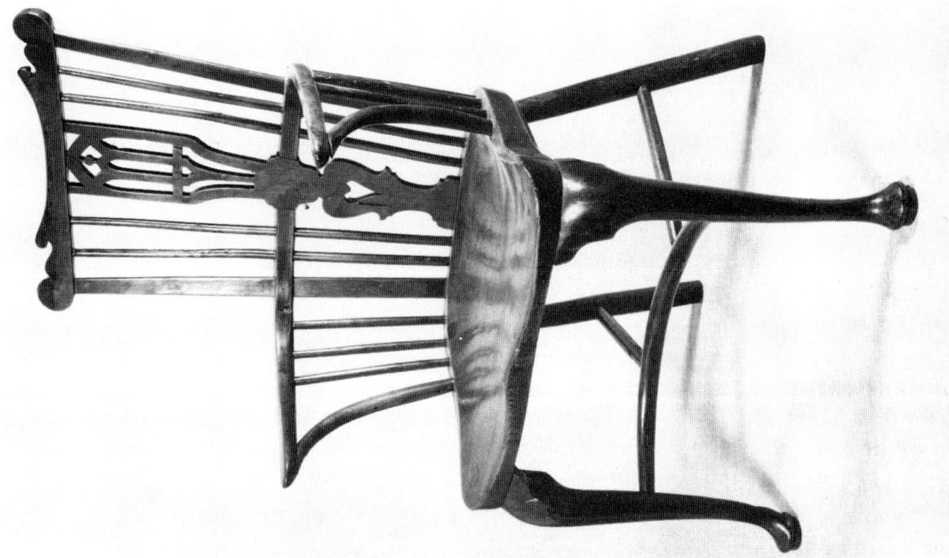

77 CHIPPENDALE YEWWOOD WINDSOR ARMCHAIR.
Victoria & Albert Museum. Crown Copyright.

78 CHIPPENDALE CARVED MAHOGANY UPHOLSTERED OPEN-ARM EASY CHAIR.
Erroneously described as *By Thomas Affleck of Philadelphia.*

79 CHIPPENDALE MAHOGANY PAW-FOOT PIECRUST TABLE.

80 CHIPPENDALE MAHOGANY TRIPOD TABLE WITH PIECRUST TOP.

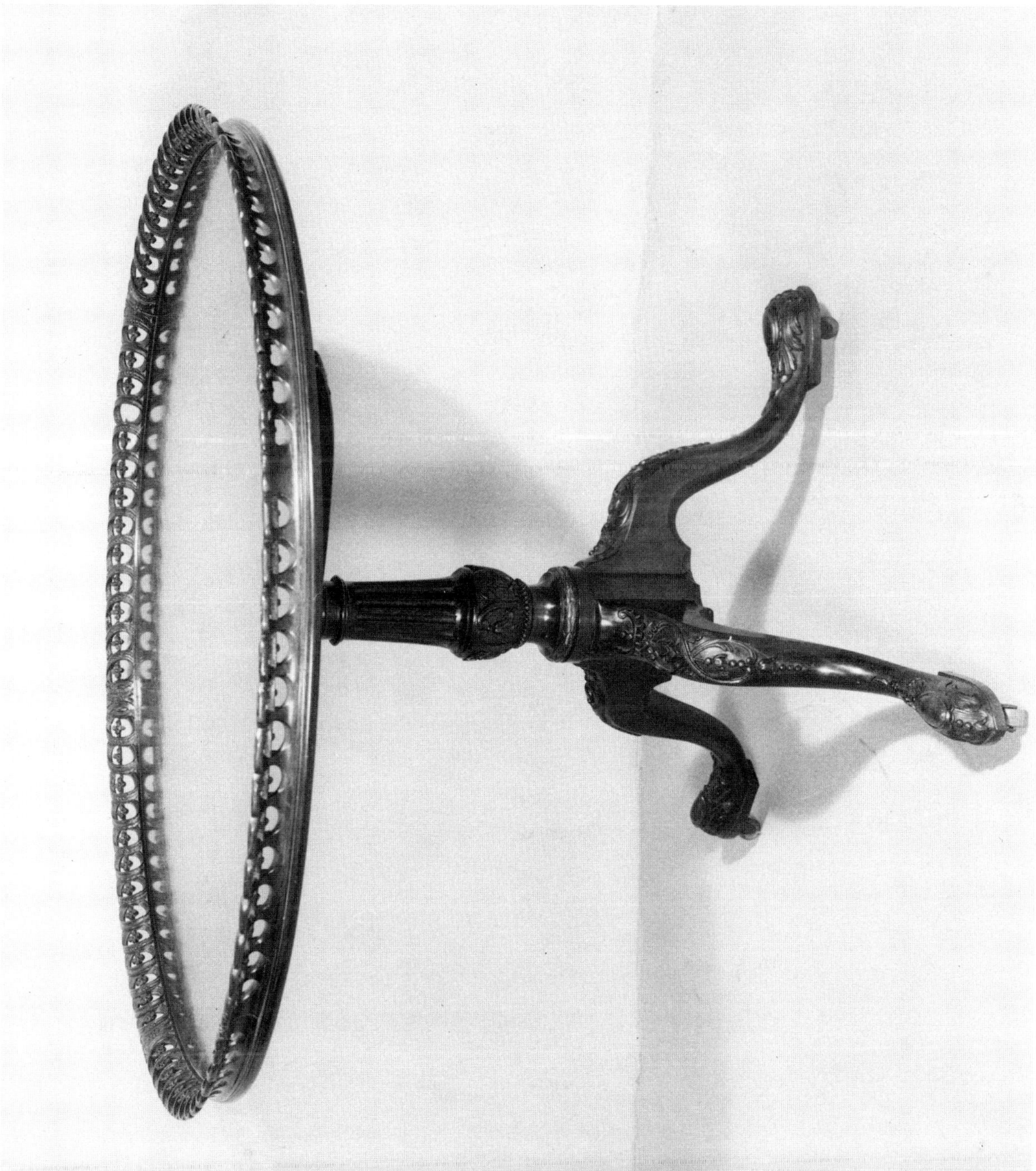

81 CHIPPENDALE MAHOGANY TRIPOD TABLE WITH PIERCED GALLERY.
Courtesy of French & Company, Inc., New York City.

82 CHIPPENDALE CARVED AND GILDED CHEVAL FIRE-SCREEN WITH IRISH NEEDLEPOINT PANEL.
Victoria & Albert Museum. Crown Copyright.

83 CHINESE-CHIPPENDALE MAHOGANY SILVER TABLE. A distinctive Dublin form of tea-china- or breakfast table with outstanding examples illustrated by Percy Macquoid, *Age of Mahogany*, Fig. 219, and by Jourdain & Rose, *English Furniture: the Georgian Period*, 69.

84 CHINESE-CHIPPENDALE MAHOGANY SILVER TABLE.

85 CHIPPENDALE MAHOGANY TRAY-TOP SILVER TABLE. Formerly believed to be *American*.

86 CHIPPENDALE FRET-CARVED MAHOGANY PEMBROKE TABLE WITH PIERCED STRETCHER. Similar tables with corresponding stretchers were produced in Pennsylvania and New England.

87 CHIPPENDALE MAHOGANY PEDESTAL WRITING TABLE. With Dublin handles of a very popular pattern. *Vide* also Percy Macquoid, *Age of Mahogany*, Fig. 161, another Dublin writing table, from *Badminton House*; and Jourdain & Rose, *English Furniture: the Georgian Period*, 5 and 83.

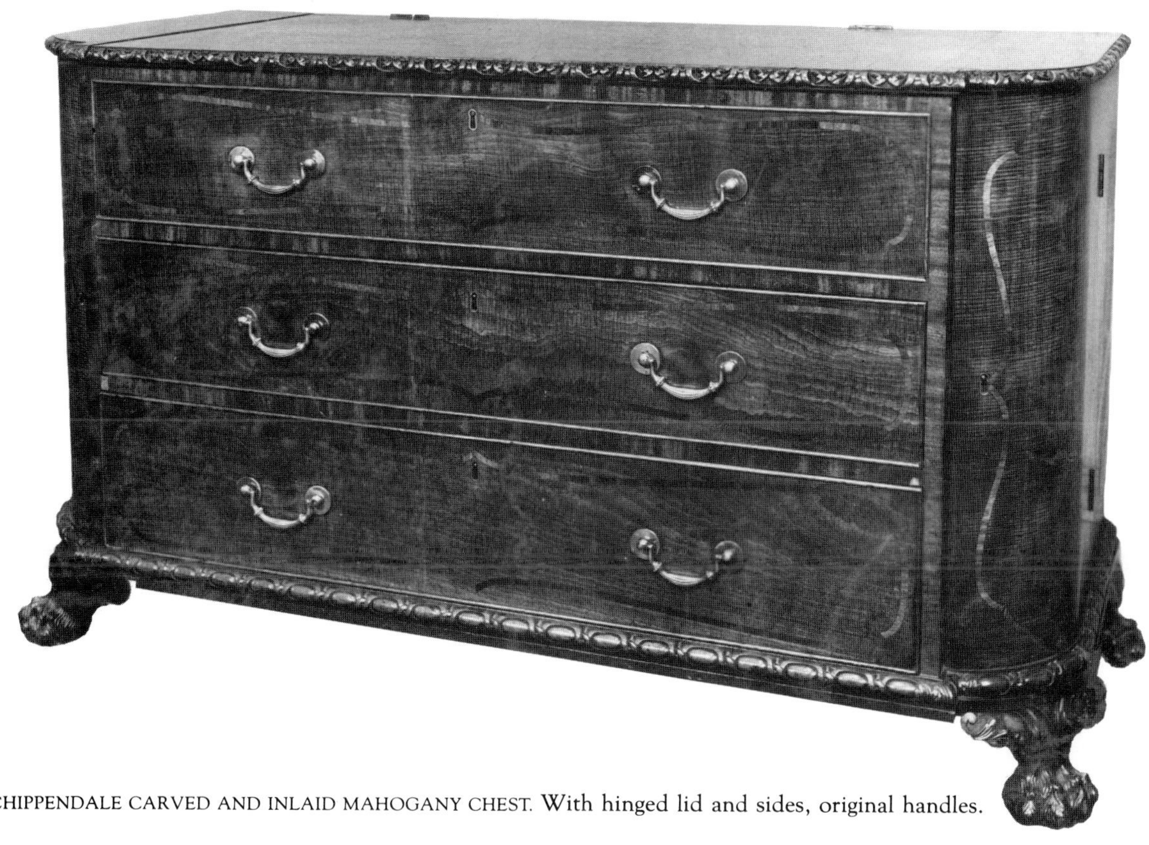

88 CHIPPENDALE CARVED AND INLAID MAHOGANY CHEST. With hinged lid and sides, original handles.

89 CHIPPENDALE CARVED AND INLAID MAHOGANY COMMODE WITH MARBLE TOP.

90 CHIPPENDALE MAHOGANY COMMODE WITH WRITING DRAWER. Handles changed.
Erroneously attributed by English authorities to Thomas Chippendale.

91 CHIPPENDALE MAHOGANY COMMODE. With original handles, the escutcheons added.
Cf. Macquoid & Edwards, *Dictionary of English Furniture*, Vol. 2, p. 132.

92 SHERATON MAHOGANY DRESSING MIRROR.
93 CHIPPENDALE MAHOGANY COMMODE. With original handles.

94 CHIPPENDALE MAHOGANY COMMODE. With original handles as in Ills. 95, 96, and in R. W. Symonds, *English Furniture From Charles II To George II*, Fig. 144.

95 LATE CHIPPENDALE MAHOGANY COMMODE.

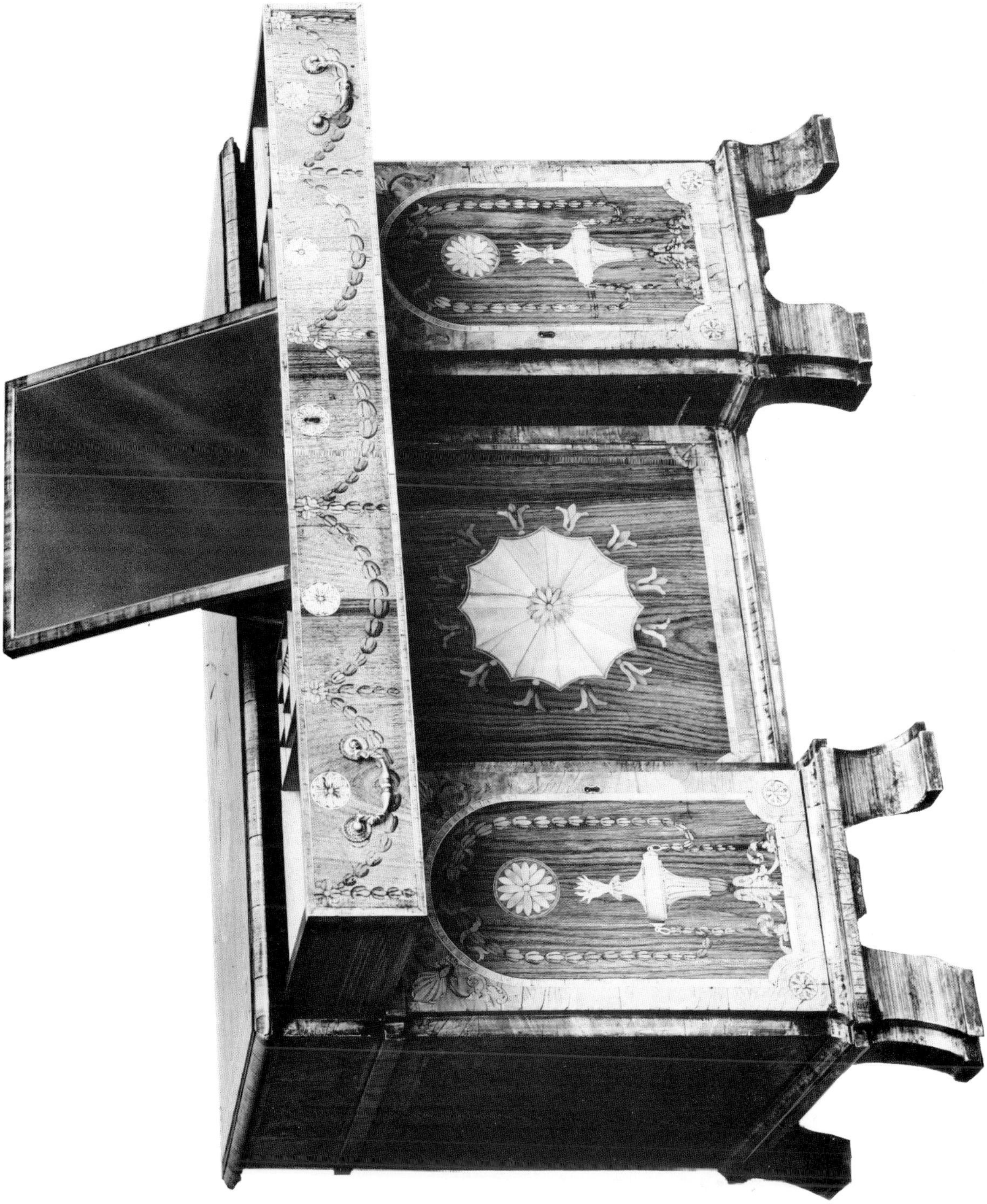

96 CHIPPENDALE ROSEWOOD MARQUETRY KNEEHOLE DRESSING COMMODE.
Victoria & Albert Museum. Crown Copyright.

97 LATE CHIPPENDALE MAHOGANY COMMODE. With original handles.

98 LATE CHIPPENDALE MAHOGANY COMMODE. With original Regency handles.

99 LATE CHIPPENDALE MAHOGANY PAGODA-TOP DISPLAY CABINET.

100 LATE CHIPPENDALE MAHOGANY CHEST-ON-CHEST. With original handles and escutcheons.

101 LATE CHIPPENDALE MAHOGANY LINEN PRESS.

102 LATE CHIPPENDALE MAHOGANY SECRETAIRE WITH GILDED CARTOUCHE FINIAL. The conventional handles original.

103 CHIPPENDALE MAHOGANY KETTLE-BASE SECRETAIRE WITH GILDED CARTOUCHE FINIAL.

104 HEPPLEWHITE MAHOGANY ARMCHAIR.

105 HEPPLEWHITE MAHOGANY SIDE CHAIR.

106 HEPPLEWHITE MAHOGANY ARMCHAIR.

107 HEPPLEWHITE MAHOGANY ARMCHAIR.

108 HEPPLEWHITE MAHOGANY UPHOLSTERED ARMCHAIR.

109 HEPPLEWHITE MAHOGANY ARMCHAIR.

110 HEPPLEWHITE MAHOGANY BERGERE.

111 HEPPLEWHITE MAHOGANY ARMCHAIR.

112 HEPPLEWHITE MAHOGANY CARD TABLE. Edged with inverse gadrooning. Victoria & Albert Museum. Crown Copyright.

113 HEPPLEWHITE MAHOGANY CARD TABLE. *En suite* with Ill. 114.

114 HEPPLEWHITE MAHOGANY UPHOLSTERED SETTEE. The seat rails centering a tablet carved with crest of the Hungerford family.

115 HEPPLEWHITE MAHOGANY SIDE CHAIR. The front legs featuring drop reeds.

116 HEPPLEWHITE MAHOGANY SIDE CHAIR.

117 HEPPLEWHITE MAHOGANY ARMCHAIR. *Vide* also *Dictionary of English Furniture*, Fig. 154, supposedly showing *Adam's influence*, and again with drop reeds.

118 HEPPLEWHITE MAHOGANY SHIELD-BACK SIDE CHAIR.

119 HEPPLEWHITE CARVED AND INLAID MAHOGANY SIDE CHAIR. Described as *American*.

120 HEPPLEWHITE MAHOGANY ARMCHAIR.

121 HEPPLEWHITE HEART-AND-SHIELD-BACK ARMCHAIR CARVED WITH SHAMROCKS. Metropolitan Museum of Art.

122 HEPPLEWHITE MAHOGANY ARMCHAIR. Carved with Prince-of-Wales plumes and inlaid with fan paterae. Victoria & Albert Museum. Crown Copyright.

123 HEPPLEWHITE INLAID MAHOGANY SIDE TABLE WITH MARQUETRY PANELS. Courtesy of Trevor, London.

124 HEPPLEWHITE AMARANTH AND HOLLY MARQUETRY DRESSING TABLE.

125 HEPPLEWHITE SYCAMORE MARQUETRY CABINET ON TABLE STAND. Distinctively inlaid with scrolling stems of rose leaves and blossoms. Victoria & Albert Museum. Crown Copyright.

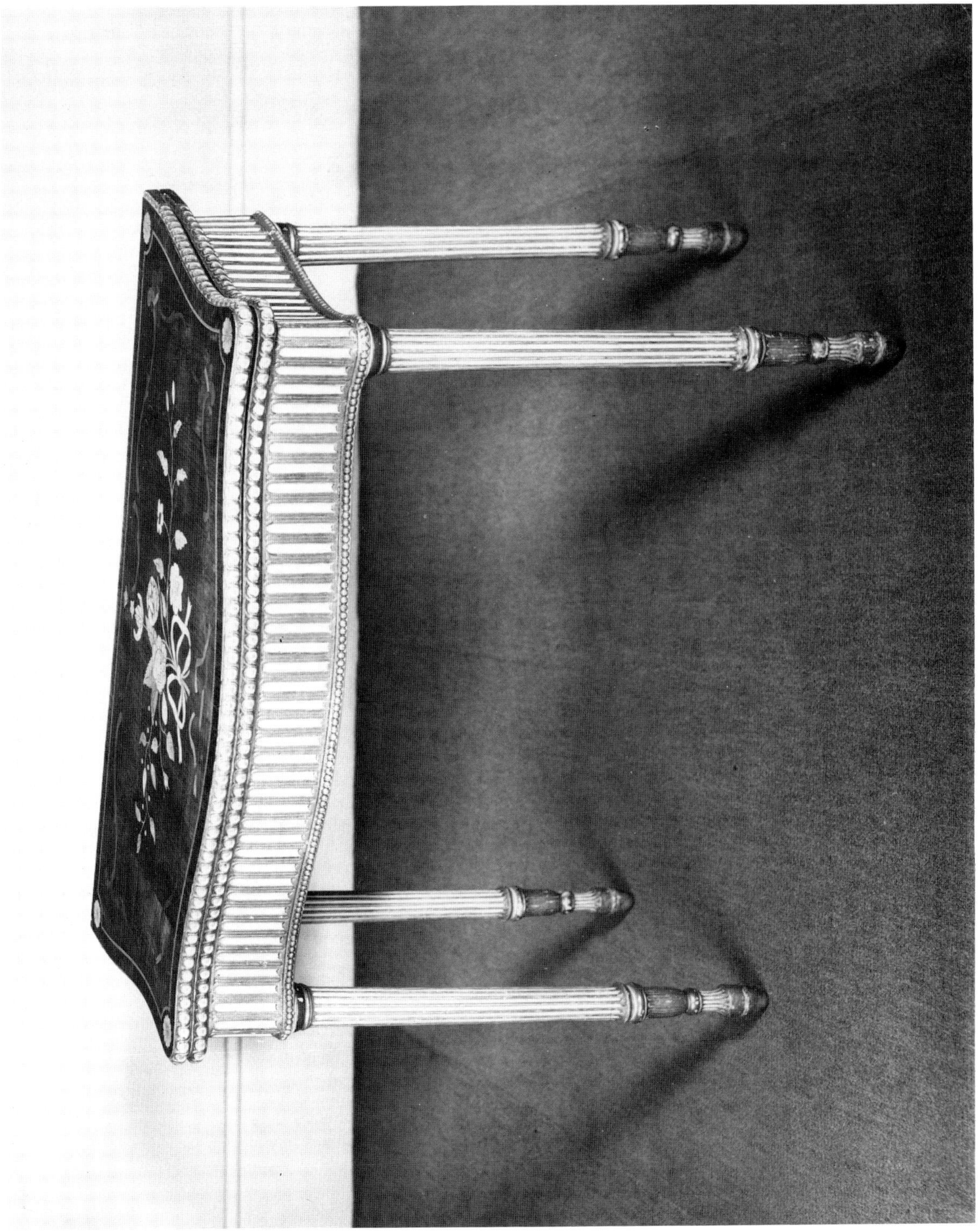

126 HEPPLEWHITE PAINTED AND PARCEL-GILDED SERPENTINE CARD TABLE. The top distinctively inlaid with scrolling stems of rose leaves and blossoms. Courtesy of Biggs of Maidenhead.

127 HEPPLEWHITE TULIPWOOD, THUJA, AND HOLLY BONHEUR DU JOUR. Mounted in ormolu; the gallery pattern continued in Sheraton and later productions. Courtesy of J. J. Wolff (Antiques) Ltd., New York City.

128 HEPPLEWHITE GILDED CONSOLE WITH INLAID MARBLE TOP. Courtesy of Asprey & Company Ltd., London.

129 "HEPPLEWHITE" MAHOGANY DRESSING COMMODE. Inlaid with distinctively checkered bandings. Victoria & Albert Museum. Crown Copyright.

130 HEPPLEWHITE SATINWOOD MARQUETRY COMMODE. Victoria & Albert Museum. Crown Copyright.

131 HEPPLEWHITE MARQUETRY COMMODE. With ormolu mounts varying from those of Ill. 130. Metropolitan Museum of Art.

132 HEPPLEWHITE MARQUETRY BREAKFRONT COMMODE. *Cf.* pattern of rose stems with Ill. 125. Ordered by the fourth Earl of Shaftesbury for St. Giles House, Dorset. Of late erroneously attributed to the London work of Thomas Chippendale. Courtesy of Christie, Manson, & Woods, London.

133 GEORGE III MAHOGANY SECRETAIRE. With original handles; *cf.* Ills. 94, 95, 96.

134 GEORGE III SOLID FIGURED YEWWOOD ARCHITECTURAL BOOKCASE.
Collection of A. Randolph Brett, High Close, Baldock, Hertfordshire.

135 GEORGE III MAHOGANY BREAKFRONT BOOKCASE. The cornice inlaid with fan paterae and mock fluting. Formerly in an old Welsh residence.

136 HEPPLEWHITE PAINTED BEECH ARMCHAIR.

137 HEPPLEWHITE MAHOGANY ARMCHAIR.

138 GEORGE III PAINTED BEECH ARMCHAIR.

139 GEORGE III DECORATED LACQUER SIDE CHAIR. The front legs featuring drop reeds.

140 SHERATON MAHOGANY SIDE CHAIR.
Courtesy of Biggs of Maidenhead.

141 SHERATON MAHOGANY SIDE CHAIR.

142 SHERATON MAHOGANY ARMCHAIR.
A back design followed in New York.

143 SHERATON MAHOGANY SIDE CHAIR.
More definitely a Dublin, rather than New York design.

144 GEORGE III YEWWOOD WINDSOR ARMCHAIR. One variation published as *Dutch*, the usual mistake in misjudging certain types of unrecognized Dublin designs. Courtesy of Mallett & Son (Antiques) Ltd., London and Geneva.

146 VICTORIAN YEWWOOD TAVERN CHAIR.

145 GEORGE III YEWWOOD WINDSOR ARMCHAIR. Featuring five Prince-of-Wales plumes, a favorite Dublin device. Courtesy of Mallett & Son (Antiques) Ltd., London and Geneva.

147 SHERATON MAHOGANY PEMBROKE TABLE.

148 SHERATON INLAID SATINWOOD WRITING TABLE.
The handles featuring Prince-of-Wales plumes.
Victoria & Albert Museum. Crown Copyright.

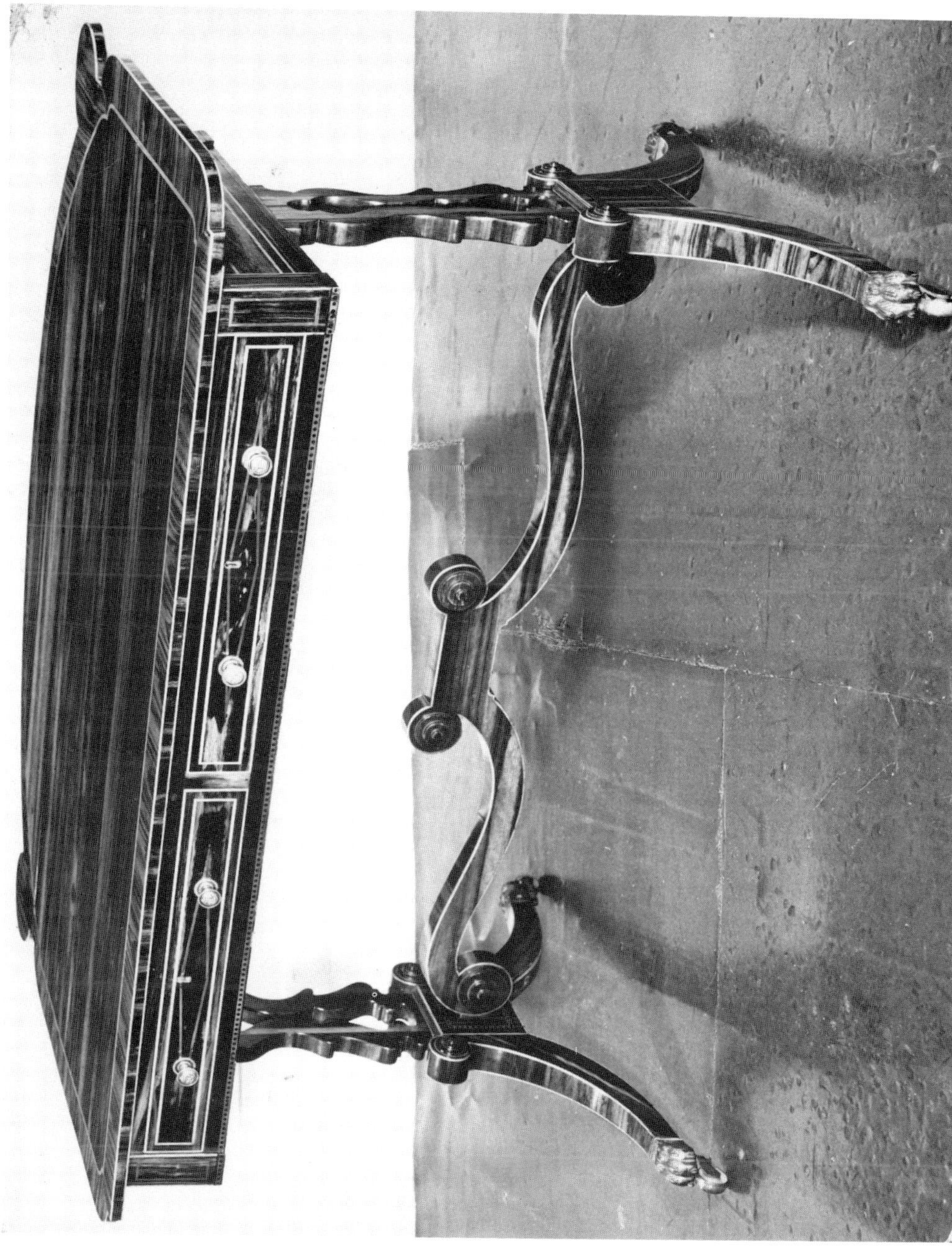

152 REGENCY CALAMANDER SOFA TABLE. With shaped drop leaf at one side.

153 REGENCY INLAID MAHOGANY CARD TABLE. Supposedly made in Salem, Massachusetts.

154 REGENCY CARVED MAHOGANY AND ORMOLU-MOUNTED CARD TABLE.

155 REGENCY BRASS-INLAID MAHOGANY AND CURLY MAPLE CARD TABLE.

156 REGENCY CARVED MAHOGANY AND ORMOLU-MOUNTED CARD TABLE.

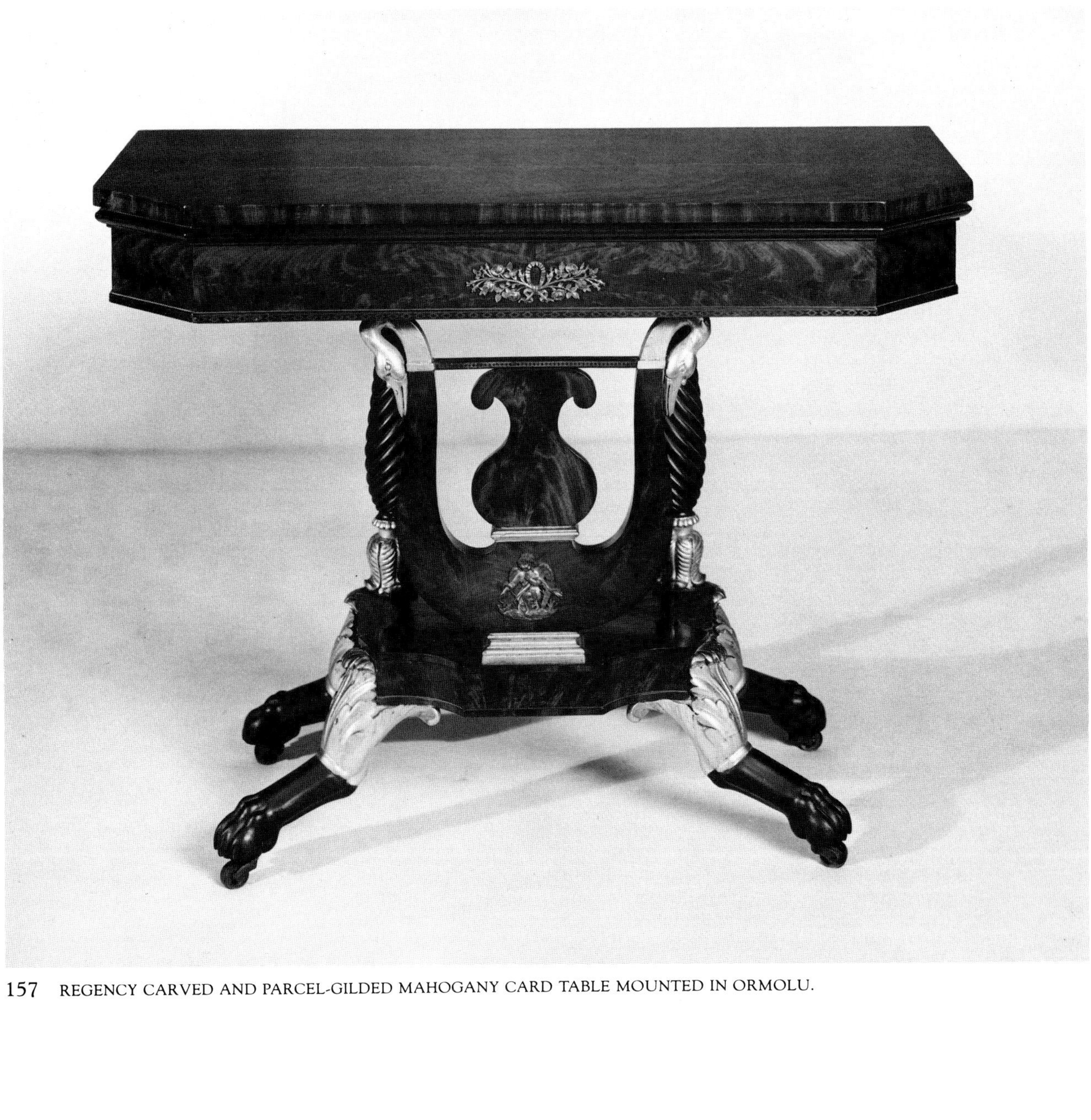

157 REGENCY CARVED AND PARCEL-GILDED MAHOGANY CARD TABLE MOUNTED IN ORMOLU.

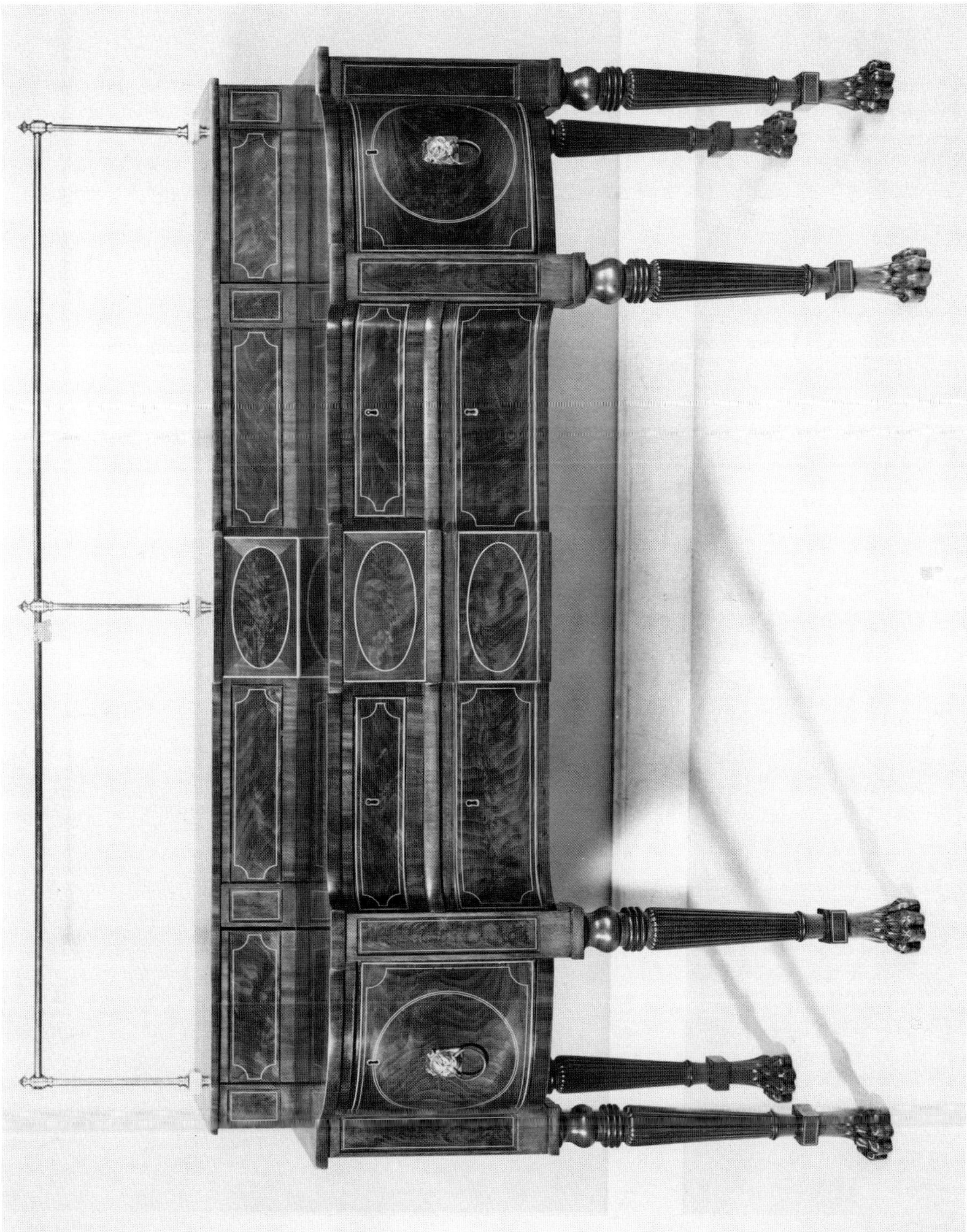

158 REGENCY INLAID MAHOGANY SIDEBOARD.

159 SHERATON MAHOGANY CHEST OF DRAWERS. Inlaid with his characteristic urns, paterae, and drapery festoons by William Moore. Original patera handles. Courtesy of French & Company, Inc., New York City.

160 SHERATON MARQUETRY COMMODE. By William Moore, the same frieze patterns of paterae flanking his distinctive two-handled urn appears in a writing table claimed as *by Robert Adam*.

161 SHERATON MAHOGANY SECRETAIRE. With original handles of a pattern which, with only slight variations, is found on Dublin cabinetwork of Queen Anne, Early Georgian and Chippendale designs. *Vide* R. W. Symonds, *English Furniture From Charles II To George II*, Fig. 131.

162 SHERATON MAHOGANY SECRETAIRE. With characteristically inlaid lower frieze of paterae and mock fluting.

163 SHERATON MAHOGANY BOOK CABINET WITH WRITING DRAWER.
With original handles and *J. Bramah Patent* lever locks supplied from London.

164 REGENCY MAHOGANY SMALL BREAKFRONT BOOKCASE.

165 SHERATON INLAID AND DECORATED SATINWOOD COMMODE. With original handles. The same type of painted festooning centering on eliptical ivory escutcheons appears in Macquoid's *Age of Satinwood*, Fig. 189.

166 SHERATON CARVED AND DECORATED SATINWOOD ARCHITECTURAL CABINET. With original handles and escutcheons of *circa* 1800. Collection of Lady Hamilton, Naples.
Courtesy of The National Gallery of Victoria, Melbourne.

167 QUEEN ANNE GILDED LOOKING GLASS WITH MIRROR-PLATE BORDER.

169 EARLY GEORGIAN VENETIAN-TYPE ÉGLOMISÉ PIER MIRROR. With carved and gilded cresting.
Victoria & Albert Museum. Crown Copyright.

168 QUEEN ANNE PIER MIRROR WITH VENETIAN-TYPE ÉGLOMISÉ DECORATION.
Victoria & Albert Museum. Crown Copyright.

170 EARLY GEORGIAN GILDED GESSO LOOKING GLASS.
Courtesy of Charles Lumb & Sons Ltd., Harrogate.

171 EARLY GEORGIAN GILDED GESSO LOOKING GLASS.

172 EARLY GEORGIAN GILDED GESSO LOOKING GLASS.

173 EARLY GEORGIAN GILDED GESSO LOOKING GLASS.
Cf. Ills. 172-3 with Edwards & Jourdain, *Georgian Cabinet-Makers*, Fig. 31, an example *Attributed to Moore and Gumley*, in *Erthig Park, Denbighshire*, North Wales.

174 EARLY GEORGIAN GILDED GESSO LOOKING GLASS.
Exhibited as *American (1725-1750)*.
Metropolitan Museum of Art.

175 EARLY GEORGIAN GILDED GESSO LOOKING GLASS.
Cf. Nutting, 2794.

176 EARLY GEORGIAN GILDED GESSO ARCHITECTURAL PIER MIRROR.

177 EARLY GEORGIAN GILDED GESSO ARCHITECTURAL PIER MIRROR.

178 EARLY GEORGIAN GILDED GESSO ARCHITECTURAL PIER MIRROR.

179 EARLY GEORGIAN GILDED GESSO ARCHITECTURAL PIER MIRROR.

180 CHIPPENDALE CARVED AND FARCEL-GILDED MAHOGANY WALL MIRROR. With ormolu candle arms. A type accepted as *American*.

181 EARLY GEORGIAN GILDED GESSO PIER MIRROR.

182 EARLY GEORGIAN CARVED MAHOGANY PIER MIRROR. With ormolu candle arms. Courtesy of Needham's Antiques, Inc., New York City.

183 EARLY GEORGIAN CARVED WALNUT AND PARCEL-GILDED OVERMANTEL MIRROR WITH OIL PAINTING.

184 EARLY GEORGIAN PARCEL-GILDED WALNUT PIER MIRROR.

185 EARLY GEORGIAN PARCEL-GILDED WALNUT PIER MIRROR.

186 EARLY GEORGIAN PARCEL-GILDED WALNUT PIER MIRROR.

187 EARLY GEORGIAN PARCEL-GILDED WALNUT PIER MIRROR.

188 EARLY GEORGIAN PARCEL-GILDED WALNUT PIER MIRROR.

189 EARLY GEORGIAN PARCEL-GILDED WALNUT PIER MIRROR.

190 EARLY GEORGIAN PARCEL-GILDED MAHOGANY PIER MIRROR.

191 EARLY GEORGIAN PARCEL-GILDED WALNUT PIER MIRROR.

192 EARLY GEORGIAN PARCEL-GILDED THUJAWOOD OVERMANTEL MIRROR WITH OIL PAINTING.
Vide also *Dictionary of English Furniture*, Fig. 43.

193 EARLY GEORGIAN PARCEL-GILDED MAHOGANY ARCHITECTURAL PIER MIRROR.

194 EARLY GEORGIAN PARCEL-GILDED MAHOGANY ARCHITECTURAL PIER MIRROR.

195 EARLY GEORGIAN PARCEL-GILDED MAHOGANY ARCHITECTURAL PIER MIRROR. Of a design similar to those distributed by John Elliott, but with varying bird finial.

196 EARLY GEORGIAN PARCEL-GILDED MAHOGANY ARCHITECTURAL PIER MIRROR. Courtesy of J. J. Wolff (Antiques) Ltd., New York City.

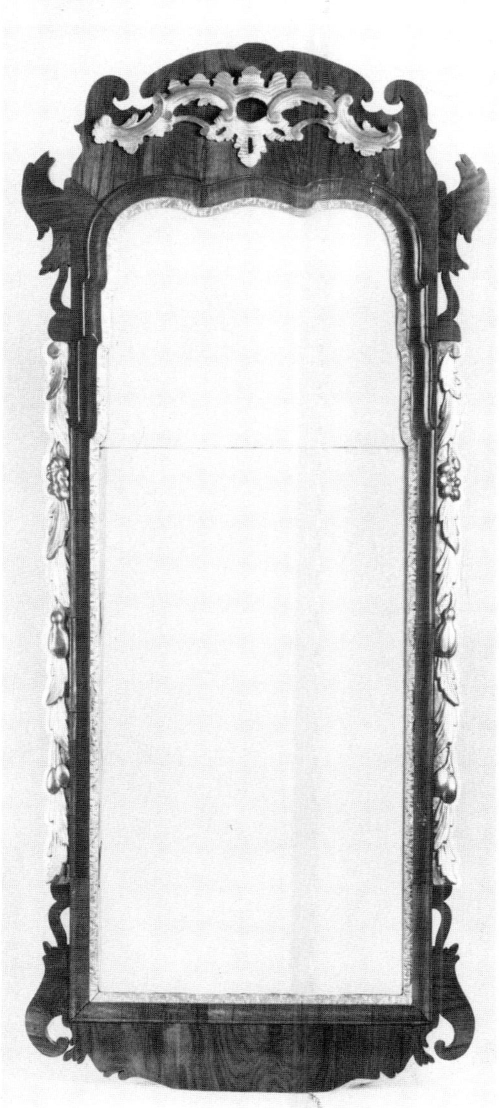

197 EARLY GEORGIAN PARCEL-GILDED WALNUT PIER MIRROR. Exhibited as *American.*

199 EARLY GEORGIAN PARCEL-GILDED WALNUT PIER MIRROR. With brass candle arms. Described as *English.*

198 EARLY GEORGIAN PARCEL-GILDED WALNUT PIER MIRROR. Described as *American or English.*

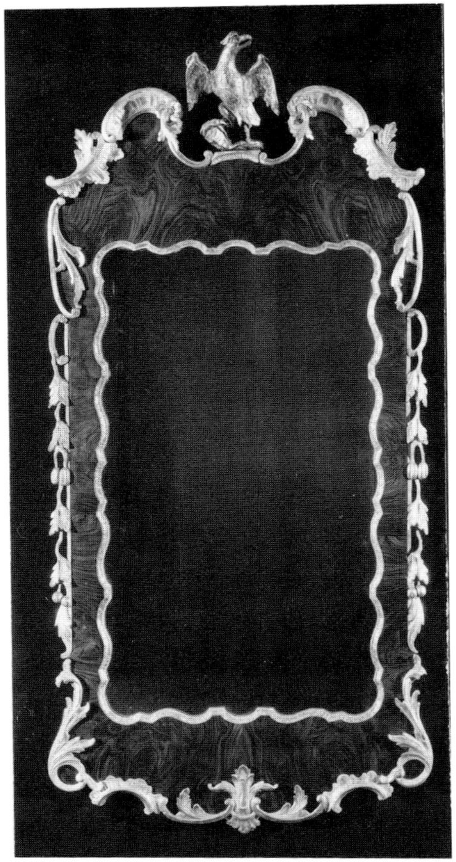

200 EARLY GEORGIAN PARCEL-GILDED WALNUT LOOKING GLASS.

201 EARLY GEORGIAN PARCEL-GILDED WALNUT ARCHITECTURAL WALL MIRROR. With bird finial as in Ill. 200.

202 CHIPPENDALE PARCEL-GILDED MAHOGANY LOOKING GLASS. Described as *American or English*.

203 EARLY GEORGIAN PARCEL-GILDED MAHOGANY PIER MIRROR. Exhibited as *American.* Metropolitan Museum of Art.

204 EARLY GEORGIAN PARCEL-GILDED WALNUT ARCHITECTURAL WALL MIRROR. Courtesy of J. J. Wolff (Antiques) Ltd., New York City.

205 EARLY GEORGIAN PARCEL-GILDED WALNUT 'CONSTITUTION' OR 'MARTHA WASHINGTON' LOOKING GLASS. Described as *American*, as *American or English*, and as *Continental.*

206 EARLY GEORGIAN PARCEL-GILDED MAHOGANY ARCHITECTURAL WALL MIRROR. With ormolu candle arms.

207 EARLY GEORGIAN PARCEL-GILDED MAHOGANY WALL MIRROR. The bird finial somewhat similar to those supplied for Elliott's importations. Courtesy of J. J. Wolff (Antiques) Ltd., New York City.

208 CHIPPENDALE PARCEL-GILDED MAHOGANY WALL MIRROR. The type distributed *circa* 1790 by William Wilmerding in New York City.

209 CHIPPENDALE PARCEL-GILDED MAHOGANY WALL MIRROR. Exhibited as *American, circa 1770.* Metropolitan Museum of Art.

210 HEPPLEWHITE PARCEL-GILDED MAHOGANY WALL MIRROR WITH ÉGLOMISÉ PANEL. The type generally claimed for New York City.

211 HEPPLEWHITE PARCEL-GILDED MAHOGANY WALL MIRROR.

212 LATE GEORGIAN FRET-CARVED AND PARCEL-GILDED MAHOGANY LOOKING GLASS. With same bird finial as in Ill. 213.
Metropolitan Museum of Art.

214 LATE GEORGIAN PARCEL-GILDED MAHOGANY LOOKING GLASS. Accepted as American, circa 1750-1775. *Cf.* eagle with that of Ill. 265. Metropolitan Museum of Art.

213 LATE GEORGIAN PARCEL-GILDED MAHOGANY LOOKING GLASS. Accepted as a Constitution mirror, American XVIII Century.

215 CHIPPENDALE CARVED AND GILDED WALL MIRROR.

216 CHIPPENDALE CARVED AND GILDED WALL MIRROR
Victoria & Albert Museum. Crown Copyright.

217 CHIPPENDALE CARVED AND GILDED WALL MIRROR.

218 CHIPPENDALE CARVED AND GILDED WALL MIRROR.
Described as American or English.

219 CHIPPENDALE CARVED AND GILDED WALL MIRROR.
Courtesy of Charles Lumb & Sons Ltd., Harrogate.

220 CHIPPENDALE CARVED AND GILDED WALL MIRROR.

221 CHIPPENDALE CARVED AND PAINTED WALL MIRROR.

222 CHIPPENDALE CARVED AND GILDED WALL MIRROR.

223 CHIPPENDALE CARVED AND GILDED WALL MIRROR. Exhibited as American in the Philadelphia Chippendale style. Metropolitan Museum of Art.

224 CHIPPENDALE CARVED AND GILDED WALL MIRROR.

225 CHIPPENDALE CARVED AND GILDED WALL MIRROR.
Exhibited as possibly of Philadelphia origin.

226 CHIPPENDALE CARVED AND GILDED WALL MIRROR.
Exhibited as a Philadelphia production.

227 CHIPPENDALE CARVED AND GILDED OVERMANTEL MIRROR.

228. CHIPPENDALE CARVED AND GILDED OVERMANTEL MIRROR. Exhibited as American. Metropolitan Museum of Art.

229 CHIPPENDALE CARVED AND GILDED OVERMANTEL MIRROR WITH OIL PAINTING.
Courtesy of Needham's Antiques, Inc., New York City.

230 CHIPPENDALE CARVED AND GILDED OVERMANTEL MIRROR.

231 CHIPPENDALE CARVED AND GILDED WALL MIRROR.

232 CHIPPENDALE CARVED AND GILDED WALL MIRROR.

233 CHIPPENDALE CARVED AND GILDED LOOKING GLASS WITH CHINESE MIRROR PAINTING.

234 CHIPPENDALE CARVED AND GILDED WALL MIRROR.

235 CHIPPENDALE CARVED AND GILDED WALL MIRROR.
Cf. Dictionary of English Furniture, Fig. 80.

236 CHIPPENDALE CARVED AND GILDED WALL MIRROR.

237 CHIPPENDALE CARVED AND GILDED OVERMANTEL MIRROR.
Vide also *Age of Mahogany*, Fig. 31; *Cf. Dictionary of English Furniture*, Fig. 70, 80.

238 CHIPPENDALE CARVED AND GILDED WALL MIRROR.

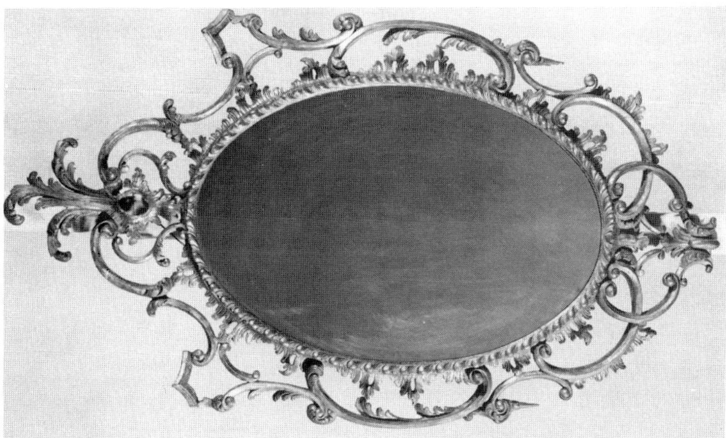

239 CHIPPENDALE CARVED AND GILDED WALL MIRROR.

240 CHIPPENDALE CARVED AND GILDED WALL MIRROR.

241 CHIPPENDALE CARVED AND GILDED WALL MIRROR.

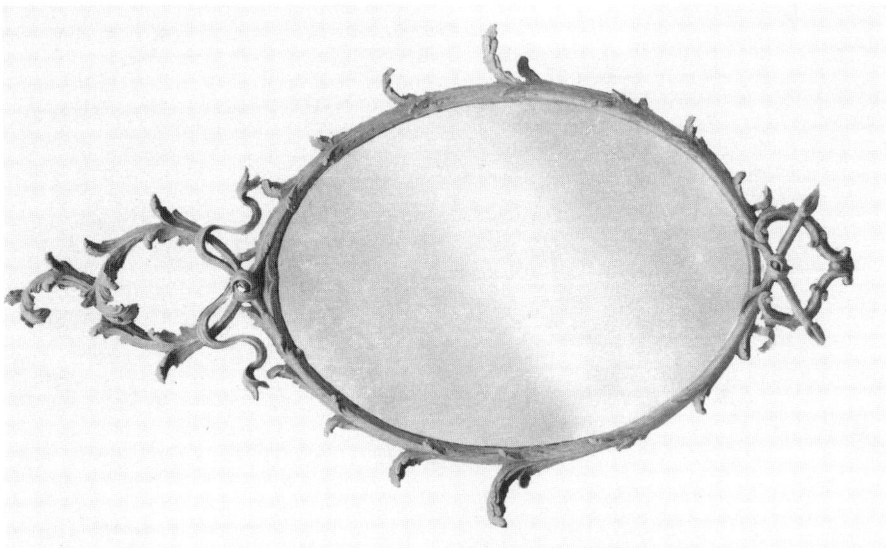

243 CHIPPENDALE CARVED AND GILDED WALL MIRROR.

242 CHIPPENDALE CARVED AND GILDED WALL MIRROR.

245 CHIPPENDALE CARVED AND GILDED WALL MIRROR.
Collection of Lord Portman, Buxted Park, Sussex.

244 CHIPPENDALE CARVED AND GILDED WALL MIRROR.
In Powerscourt, County Wicklow.
Photograph courtesy of the late Noel C. Hartnell, Dublin.

246 ADAM CARVED WOOD AND GESSO GILDED OVAL WALL MIRROR.

247 ADAM CARVED AND GILDED OVAL WALL MIRROR.
Courtesy of Needham's Antiques, Inc., New York City.

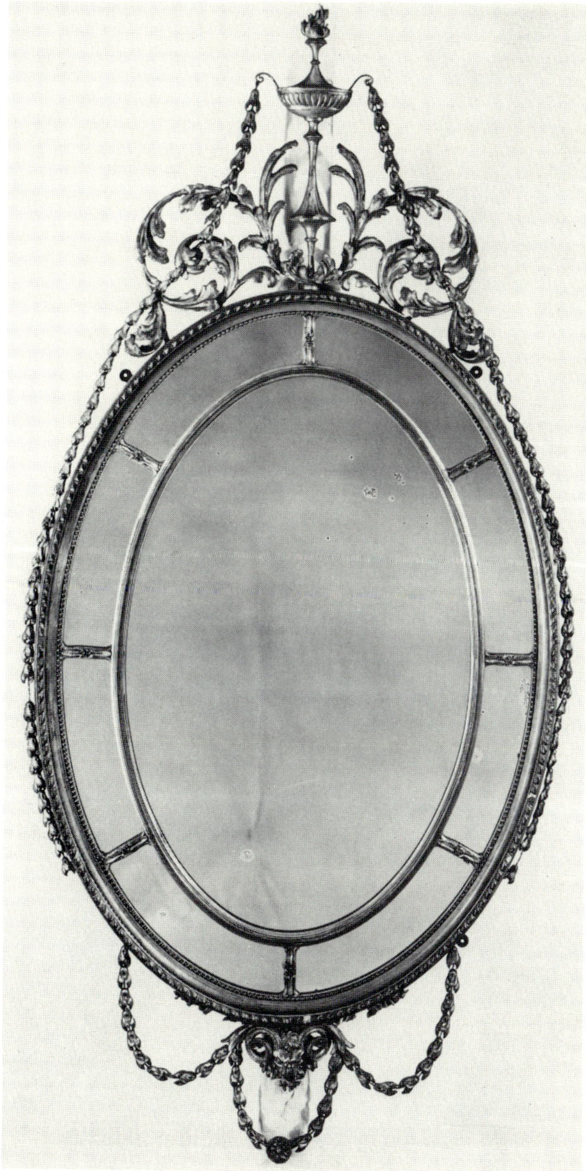

248 ADAM GILDED OVAL WALL MIRROR.

249 ADAM GILDED OVAL WALL MIRROR WITH ÉGLOMISÉ MEDALLION.

250 ADAM GILDED OVAL WALL MIRROR.

251 ADAM GILDED OVAL WALL MIRROR.

252 ADAM-HEPPLEWHITE GILDED GIRANDOLE.

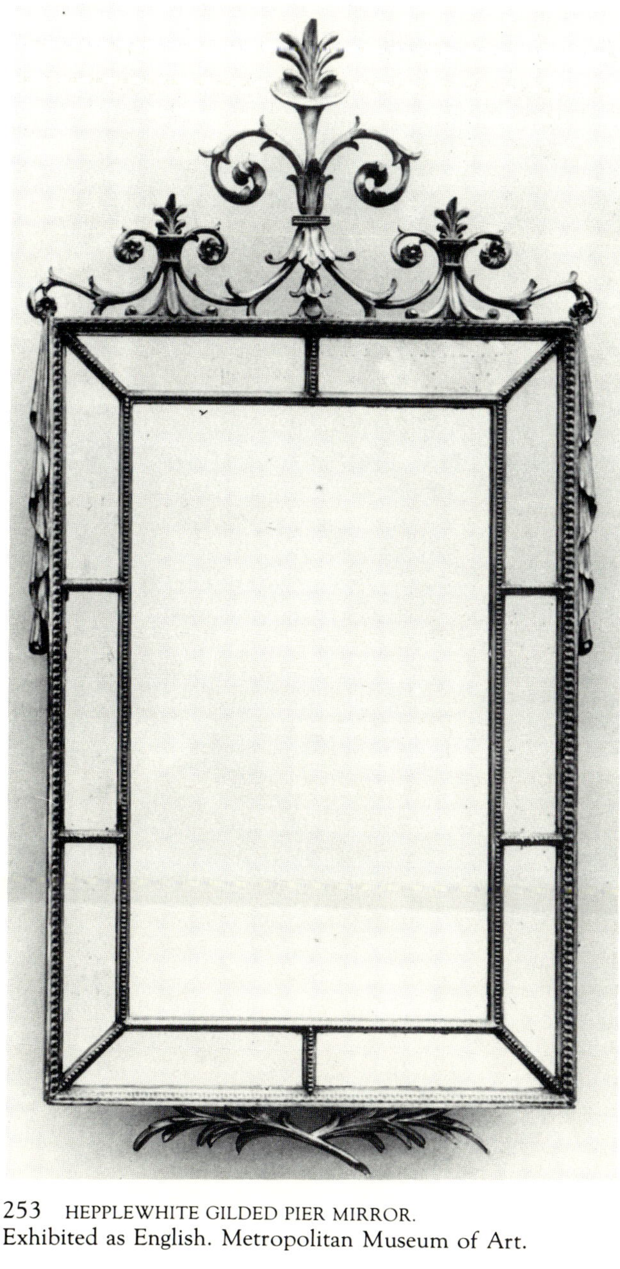

253 HEPPLEWHITE GILDED PIER MIRROR.
Exhibited as English. Metropolitan Museum of Art.

254 HEPPLEWHITE GILDED WALL MIRROR.
Exhibited as English. Metropolitan Museum of Art.

255 LATE GEORGIAN GILDED WALL MIRROR.

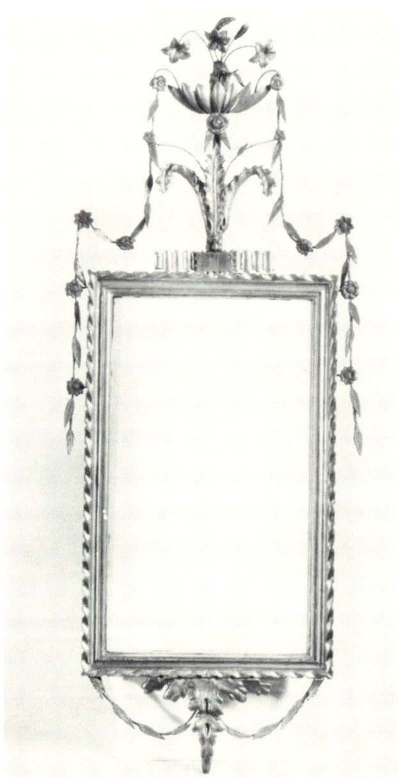

256 HEPPLEWHITE GILDED WALL MIRROR. Sold as American, *circa* 1780. Collection of Louis G. Myers, American Art Association, 1921.

257 HEPPLEWHITE GILDED WALL MIRROR.
Sold as American or English.

258 SHERATON GILDED EAGLE WALL MIRROR WITH ÉGLOMISÉ PANEL. Sold as CARVED, though with gesso ornamentation; and as with original, though fraudulently applied, label of *Cermenati & Monfrino, Boston, Massachusetts.*

259 SHERATON GILDED WALL MIRROR WITH ÉGLOMISÉ PANEL. As in Ill. 258, the églomisé landscape typical of the Irish school of decorative painting, *circa* 1810. Sold as English.

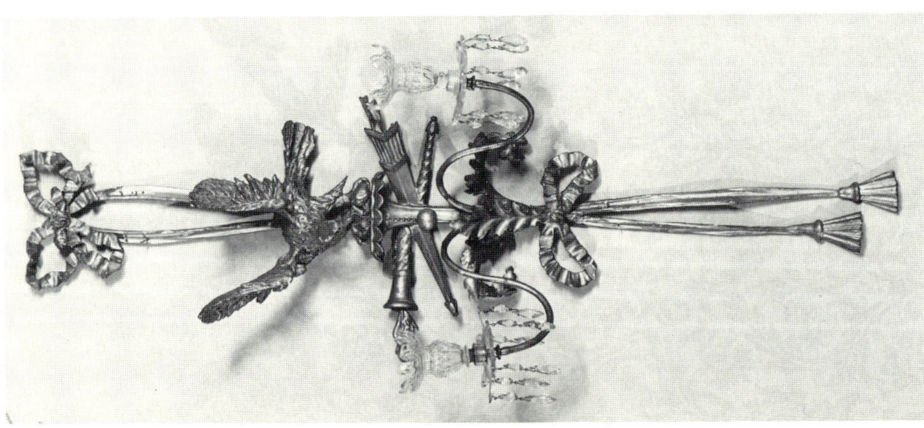

260 GEORGE III GILDED EAGLE GIRANDOLE.

261 SHERATON GILDED WALL MIRROR WITH ÉGLOMISÉ PANEL.

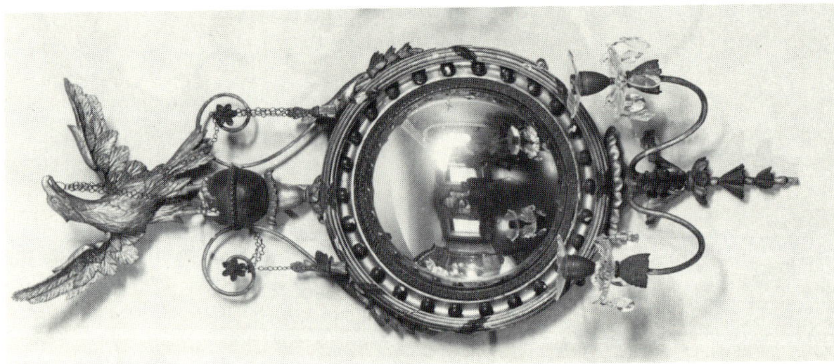

262 REGENCY GILDED CONVEX EAGLE GIRANDOLE MIRROR.

263 REGENCY GILDED CONVEX EAGLE GIRANDOLE MIRROR.

264 REGENCY GILDED CONVEX EAGLE GIRANDOLE MIRROR.

265 REGENCY GILDED AND EBONIZED CONVEX EAGLE-AND-DOLPHIN GIRANDOLE MIRROR.
Sold as American or English.

266 LATE REGENCY GILDED CONVEX EAGLE MIRROR.
Matching one in Moore Abbey, County Kildare.

267 REGENCY ORMOLU AND SAPPHIRE-BLUE ENAMEL CHANDELIERE.

268 SALON IN POWERSCOURT, COUNTY WICKLOW. Photograph courtesy of the late Noel C. Hartnell, Dublin.

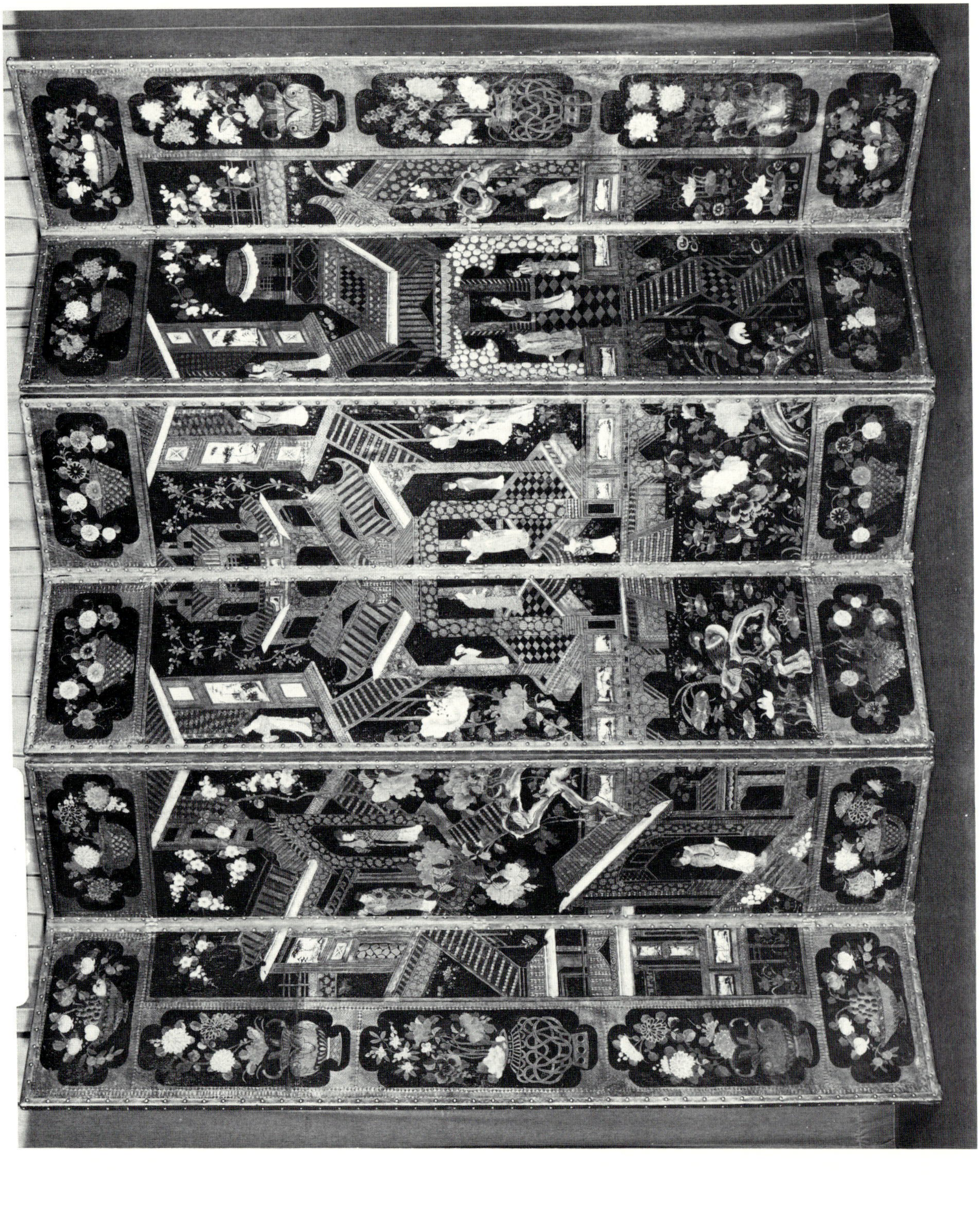

269 GEORGE II PAINTED AND GILDED LEATHER SIX-FOLD SCREEN.

Index

[Italic figures refer to illustrations]